Series/Number 07-101

INTERPRETING PROBABILITY MODELS
Logit, Probit, and Other Generalized Linear Models

TIM FUTING LIAO
University of Illinois, Urbana-Champaign

SAGE PUBLICATIONS
International Educational and Professional Publisher
Thousand Oaks London New Delhi

For information address:

SAGE Publications, Inc.
2455 Teller Road
Thousand Oaks, California 91320

SAGE Publications Ltd.
6 Bonhill Street
London EC2A 4PU
United Kingdom

SAGE Publications India Pvt. Ltd.
M-32 Market
Greater Kailash I
New Delhi 110 048 India

Printed in the United States of America

Library of Congress Catalog Card No. 89-043409

Liao, Tim Futing.
    Interpreting probability models: logit, probit, and other
generalized linear models / Tim Futing Liao.
        p.   cm.—(A Sage university papers series. Quantitative
applications in the social sciences; no. 07-101)
    Includes bibliographical references.
    ISBN 0-8039-4999-5 (pbk.)
    1. Linear models (Statistics)   2. Logits.   3. Probits.   I. Title.
    II. Series.
    QA279.L52   1994
    519.5′38—dc20   94-13978

94 95 96 97 98   10 9 8 7 6 5 4 3 2 1

Sage Production Editor: Astrid Virding

When citing a university paper, please use the proper form. Remember to cite the current Sage University Paper series title and include the paper number. One of the following formats can be adapted (depending on the style manual used):

(1) LIAO, T. F. (1994) *Interpreting Probability Models: Logit, Probit, and Other Generalized Linear Models.* Sage University Paper series on Quantitative Applications in the Social Sciences, 07-101. Thousand Oaks, CA: Sage.

OR

(2) Liao, T. F. (1994). *Interpreting probability models: Logit, probit, and other generalized linear models* (Sage University Paper series on Quantitative Applications in the Social Sciences, series no. 07-101). Thousand Oaks, CA: Sage.

CONTENTS

SERIES EDITOR'S INTRODUCTION

What is the probability that something will occur? How is that probability altered by a change in some independent variable? These are questions research workers not infrequently ask. Take a common political science example, with data from a sample survey of voters. The dependent variable (Y) of interest is vote choice between two political parties, Liberal (scored "0") and Conservative (scored "1"). Does the probability of a Conservative vote increase with increasing income (independent variable X, measured in 1,000 dollar units)? One answer comes from the linear probability model, where ordinary least squares (OLS) regression is applied, yielding the following results:

$$Y = -.02 + .01X + e.$$

According to the linear probability model interpretation, the slope estimate (.01) suggests that a unit increase (1,000 dollars) in income adds, on average, an additional point to the probability of voting Conservative. Also, we see that the predicted probability of a Conservative vote for those with $X = 30,000$ dollars would be .28. Although tidy, these OLS results lack certain desirable estimator properties. Because of the binary nature of the dependent variable, the error term is necessarily heteroscedastic, meaning the estimates are inefficient. Further, this model generates certain nonsense predictions, with probability estimates outside (0,1). For instance, for voters with no income, predicted probability $= -.02$; for voters with 103,000 dollars in income, predicted probability $= 1.01$. Finally, the true relationship between vote and income may well be curvilinear, with unit changes at the extremes of income having less effect, meaning that this OLS slope estimate is biased.

Therefore, even though the linear probability model from OLS has initial appeal, it should generally be avoided in interpreting probabilities. In its stead, a more appropriate technique should be selected, from among those carefully explicated by Professor Liao. With regard to the example at hand, a better choice would be the logit model, where the dependent variable is transformed thusly, and maximum-likelihood estimation conducted,

$$\log\left[\frac{\text{Prob}(y=1)}{1-\text{Prob}(y=1)}\right] = c + dX + v.$$

For the probability models, Professor Liao spells out four basic interpretations. He indicates that the most useful interpretation for the logit model is that of odds and odds ratios. To illustrate, he explores data from the National Survey of Children, constructing a logit model of adolescent sexual activity. By exponentiating (i.e., taking the antilog with the base e) the logit coefficients, he is able to conclude, for example, that the odds for adolescent males having sexual intercourse are about 1.9 times higher than females. A second common way to interpret logit results is to predict the probability of an event, based on a certain set of X values. To continue Professor Liao's illustration, it can be predicted that, given the adolescent respondent is white and female, the probability of intercourse is 0.146.

The logit model is but one of several covered by Professor Liao. He also evaluates its leading alternative—the probit model. The cumulative probability functions of the probit and logit models are quite similar, so they usually generate predicted probabilities that are almost identical. Logit, however, has the advantage that these predicted probabilities can be arrived at by hand calculator. Further, when there are many observations at the extremes of the distribution, then logit is preferred over probit.

Although much logit and probit modeling has concentrated on binary dependent variables, this is not a necessary limitation. After explicating sequential logit and probit models, Professor Liao explains how to interpret ordinal logit and probit models. He goes on to multinomial logit models, where no ordinality can be imposed on the dependent variable categories. (Interestingly, he explains that computational difficulties make a "multinomial" probit a rarity.) The model discussions end with Poisson regression, used to estimate the probability of rare events. Professor Liao reminds us that this is yet another situation in which OLS regression is mistakenly used. Because of its comprehensiveness, and its unifying theme of interpretation in probability terms, this monograph is a logical one to read after mastering single-equation regression methods.

—*Michael S. Lewis-Beck*
Series Editor

ACKNOWLEDGMENTS

I would like to thank Karen Folk, John Mirowsky, Wei Wu, and two anonymous referees for their valuable comments in the process of preparing this monograph.

INTERPRETING PROBABILITY MODELS
Logit, Probit, and Other Generalized Linear Models

TIM FUTING LIAO
University of Illinois, Urbana-Champaign

1. INTRODUCTION

Social scientists use probability models to answer a variety of substantive questions. The interpretation of these models is often ignored and is confusing to many. The purpose of this monograph is to introduce a systematic way for interpreting a variety of probability models commonly used by social scientists. To further illustrate the importance of probability models and their interpretation, I answer two questions in the introduction: (1) Why probability models? and (2) Why interpretation?

Why Probability Models?

A basic statistical method in the social scientist's toolbox is linear (or linearizable) regression analysis, which requires a continuous dependent variable. Much of what social scientists study, however, cannot be analyzed with the classical regression model, because many attitudes, behaviors, characteristics, decisions, and events in social science research—be they intrinsically continuous or not—are measured in discrete, nominal, ordinal, or, in short, noncontinuous ways.

A number of statistical models available for analyzing such data are normally presented and discussed in relation to the type of data, such as "binary data analysis," "ordinal data analysis," "categorical data analysis," or "discrete choice analysis," or as a specific model, for example, logit or probit models. One common characteristic of these related statistical methods is that they all model the probability of an event—how likely the event is to occur. Therefore, in this monograph I refer to all statistical

1

models analyzing event probability by a common name: probability models. The probability models discussed include binary, sequential, and ordinal logit and probit; multinomial logit; conditional logit; and Poisson regression models.

Why Interpretation?

In the past decade or so a number of comprehensive books have been published in the economic, statistical, and social sciences on probability models. A cursory examination of the references I often use for teaching and research produces the following list: Agresti's *Analysis of Ordinal Categorical Data* (1984) and *Categorical Data Analysis* (1990), Aldrich and Nelson's *Linear Probability, Logit, and Probit Models* (1984), Ben-Akiva and Lerman's *Discrete Choice Analysis: Theory and Application to Travel Demand* (1985), Bishop, Fienberg, and Holland's *Discrete Multivariate Analysis: Theory and Practice* (1975), Cox and Snell's *The Analysis of Binary Data* (1989), DeMaris's *Logit Modeling* (1992), Dobson's *An Introduction to Generalized Linear Models* (1990), Maddala's *Limited Dependent and Qualitative Variables in Econometrics* (1983), McCullagh and Nelder's *Generalized Linear Models* (1989), Santer and Duffy's *The Statistical Analysis of Discrete Data* (1989), Train's *Qualitative Choice Analysis: Theory, Econometrics, and an Application to Automobile Demand* (1986), and Wrigley's *Categorical Data Analysis for Geographers and Environmental Scientists* (1985). Among these, Aldrich and Nelson's and DeMaris's are Sage University Papers, and others are more extensive monographs. Probability models are also dealt with to a varying extent in general statistical or econometric texts such as Hanushek and Jackson's *Statistical Methods for Social Scientists* (1977) and Greene's *Econometric Analysis* (1990; or its 1993 second edition). In addition, there are excellent review essays on qualitative response models by Amemiya (1981) and by McFadden (1976, 1982).

Most of the above are comprehensive treatments that include model specification, model estimation, model selection, hypothesis testing, and assessment of overall model fit. They offer little advice on how to interpret coefficient estimates, much less interpret them in a systematic manner. Some social scientists have misgivings about these probability models precisely because of the difficulty with interpretation. As a result, they may shy away from probability models and opt for more familiar though perhaps inappropriate methods such as linear regression.

The objective of this monograph is to demonstrate how to interpret results from various probability models. Both novice and expert users of these models will benefit if interpretation is their major shortcoming as well as interest. Readers who are interested in issues other than interpretation, such as model estimation and selection, should refer to the literature given above.

The following chapter reviews generalized linear models, under which all the probability models discussed in the monograph can be subsumed. The chapter then presents a systematic way for interpreting results from generalized linear models. The same methods of interpretation are carried through from Chapters 3 through 8, with modifications for each type of probability model: Chapter 3, binary logit and probit models; Chapter 4, sequential logit and probit models; Chapter 5, ordinal logit and probit models; Chapter 6, multinomial logit models; Chapter 7, conditional logit models; and Chapter 8, Poisson regression models. In the concluding chapter a brief summary of the ways of interpreting probability models is given, the features of some general texts on these models are reviewed, and some further comments are made on interpreting parameter estimates from probability models.

2. GENERALIZED LINEAR MODELS AND THE INTERPRETATION OF PARAMETERS

In this chapter I review generalized linear models and establish a systematic method of interpreting parameter estimates based on a generalized linear models framework. These models extend classical linear models, and all the probability models to be discussed can be subsumed under generalized linear models. Therefore, specific treatments, such as interpretation, of the models belonging to this family can also be generalized and regarded as common to all models of the category.

Generalized Linear Models

The presentation here follows and simplifies Dobson's (1990) and McCullagh and Nelder's (1989) exposition. The ith observation y_i is assumed to be a realization of a random variable Y_i whose expected values are given by $E(Y_i) = \mu_i$. For parsimony in presentation, we drop the subscript i because the vector of observations is understood. When we study random variable Y using a linear model, we specify its expectation as a linear combination of K (with the subscript k running from 1 to K) unknown parameters and covariates (or explanatory or independent variables):

$$E(Y) = \mu = \sum_{k=1}^{K} \beta_k x_k .$$ [2.1]

This is the ordinary linear model, and the equation should remind the reader of a linear regression model. To create a more generalizable model, we introduce the variable, η, which links the function $\sum_k \beta_k x_k$ to μ, although not necessarily in a *linear* fashion.

In general, we specify η as a linear predictor produced by $x_1, x_2, \ldots, x_K$. Regardless of the type of model, the set of explanatory variables always linearly produce η, which is a predictor of Y. The relation between η and the x variables is given by

$$\eta = \sum_{k=1}^{K} \beta_k x_k .$$ [2.2]

The function of the relation between η and μ, however, is to be specified. The link between η and μ distinguishes one member of the generalized linear models from another. There are many possible link functions, $\eta = g(\mu)$, between η and μ. In the following, however, we focus only on those relevant to the models in the book.

(1) Linear:

$$\eta = \mu .$$

It is apparent from Equations 2.1 and 2.2 that the link in classical linear models is identity.

(2) Logit:

$$\eta = \log[\mu/(1-\mu)] .$$

Applying this link function to Equation 2.2, we specify a logit model that takes a binary outcome variable.

(3) Probit:

$$\eta = \Phi^{-1}(\mu) \ ,$$

where Φ^{-1} is the inverse of the standard normal cumulative distribution function. Similarly, this link function specifies a probit model that examines a binary outcome variable.

(4) Logarithm:

$$\eta = \log\mu \ .$$

Typically, natural logarithm is used, and a Poisson regression model is specified.

(5) Multinomial logit:

$$\eta_j = \log(\mu_j/\mu_J) \ ,$$

where j indicates the jth in $1, \ldots, J$ response categories. This link function is a natural extension of link function (2), in which J is equal to 2, and gives a multinomial logit model in which a polytomous outcome variable is studied.[1]

The choice of link function, or statistical model, to use depends on the distribution of the data and theory, which enables the researcher to understand the nature of the data. Specifically, the distribution of the random component in Y (the part that cannot be systematically explained by x variables) determines the link function and the type of generalized linear model. In generalized linear models, the distribution of the random component comes from an exponential family, in which the normal, the binomial, the Poisson distribution, and some others all belong (see McCullagh & Nelder 1989). When this distribution is normal, the linear link function arises. All applications of ordinary least squares (OLS) regression have a normality assumption on the random component of Y, which implies the continuity of Y.

The link functions (2) and (3) are both based on the binomial distribution. We have many variables in the social sciences that follow this distribution. Binary outcomes such as voted or not voted, dead or living, agreed or disagreed, migrated or not migrated, and the general occurrence or nonoccurrence of an event all generate the binomial distribution. Logit and probit models are often used to study these events.

The logarithm link function follows the Poisson distribution. The Poisson distribution is often assumed, so that the Poisson regression may be used, in the social sciences to study count variables such as presidential vetoes, accidents, number of visits to a physician or a dentist per year, and similar rare events.

Link function (5) assumes a multinomial distribution. Such a distribution is generated by outcomes with more than two choices, such as choice of a presidential candidate, modes of commuting to work, contraceptive choice, migration destination, or selection of a college major. Multinomial logit models are used to study these polytomous events. It is possible, however, that our theory will shape our understanding of the nature of the data and thus determine the link function to use. For instance, in studying a three-candidate presidential election such as the U.S. 1992 election, one researcher may perceive both Clinton and Perot as forces for change and Bush as maintaining the status quo, thereby believing in a binomial distribution of the data and using a binary logit or probit model. Another researcher may see all three candidates as politically unique, thereby choosing a multinomial distribution and a multinomial logit model.

Interpretation of Parameter Estimates

From Equation 2.2 we know that the effect of each x on η is always linear. Therefore, interpreting parameter estimates as linear effects on the predictor η must be common to all generalized linear models. Such interpretation, however, may not be intuitively appealing. For logit and probit models, the linear effect given by a parameter estimate on η suggests the effect of the corresponding x on the logit $\{\log[\mu/(1-\mu)]\}$ and probit $[\Phi^{-1}(\mu)]$, respectively. Few people think in terms of a logit or probit function. Fortunately, interpreting the linear effect is only one of several possible ways of making sense of parameter estimates from probability models. In the remainder of this section I describe a systematic method of interpreting parameter estimates from these models.

One characteristic common to all generalized linear models is that each estimate gives the partial effect of a coefficient with the effects of other x variables being controlled. No matter which probability model we interpret, we will always need to keep in mind the condition of ceteris paribus or other things being equal, similar to the way we interpret linear regression estimates. This point applies to all generalized linear models; thus, we will not reiterate it for every probability model.

Speaking from the perspective of generalized linear models, I identify five ways of interpreting parameter estimates from probability models. They are the interpretation of the sign of parameter estimates and their statistical significance, predicted values of η or transformed η given a set of values in the explanatory variables, the marginal effect of an explanatory variable on η or transformed η, the predicted or forecast probabilities given a set of values in the explanatory variables, and the marginal effect of an explanatory variable on the probability of an event. The reader will notice in Chapters 3 through 8 that the interpretation of each probability model follows closely that of the generalized linear model, and that the only differences are due to the individual link functions and model specifications.

1. The Sign of Parameter Estimates and Their Statistical Significance. This simple method of interpretation can be applied to any probability model because it ignores the link function in generalized linear models. Many empirical social science researchers do not go beyond this level of interpretation when they use probability models, partly because of the difficulty with interpretation in these models and partly because this practice is common in OLS regression, as well. Data analysts typically examine the sign (+ or −) of coefficient estimates and their significance test.

Given a significant statistical test, a positive sign of a parameter estimate suggests the likelihood of the response (event) increases with the level or presence of x, with the other xs held constant, depending on whether the variable is continuous or dichotomous. Conversely, a negative sign of the estimate suggests that the likelihood of the response (event) decreases with the level or presence of x. Such interpretation is vague because we do not know how much x increases or decreases the likelihood of the response (or event) or what the functional form of such an effect is. It is, however, easy to apply, because of its immunity to the link function and thus to differences among probability models. Due to its simplicity, we will not discuss the interpretation in terms of sign and significance in later chapters.

An insignificant test at a conventional α level such as 0.05 suggests that the effect of an x on the response variable is not statistically different from zero, given that there exists little multicollinearity between this and the other xs in the model. Some researchers comment on whether the sign of nonsignificant estimates conforms to the expected direction of effects. Such comment may not be meaningful because of the insignificant test.

2. Predicted Values of η or Transformed η Given a Set of Values in the Explanatory Variables. Although it is straightforward to predict log-odds or odds in the logit or Z scores in the probit model, such predictions are not widely used and thus not discussed here for the sake of space. The predicted transformed η, however, can be quite practical in the Poisson regression model because that gives predicted y, the event count (see Chapter 8).

3. Marginal Effect on η or Transformed η. Regardless of the specific probability model, the estimated β_k gives a marginal effect of the corresponding x_k on η. Equivalently, we may say that the estimate gives the marginal effect of x_k on the link function. Just as with a classical regression model, these marginal effects are linear; thus, their interpretations are straightforward. As with an ordinary linear model, we can say that a unit increase in x_k would incur an estimated β_k amount of increase (or decrease) in η, other things being equal.

However, such an interpretation of the marginal effect, though straightforward, is not all that useful because η is not directly interpretable (or Y is not interpretable in terms of η) unless the link is identity, as in the case of the classical linear model. Often, it is more meaningful to interpret the effect of an estimated β_k on a transformed η rather than on η itself. How to transform η depends on the link function. Take the logit link function, for example, $\eta = \log[\mu/(1 - \mu)]$. If we exponentiate both sides, what remains on the right-hand side will be an odds, $\exp(\eta) = \mu/(1 - \mu)$. Thus, the effect of x_k as represented by $\exp(\beta_k)$ now is on the odds rather than on the logit or log-odds. The interpretation becomes intuitively appealing because odds can be encountered in everyday life events such as horse-racing and sweepstakes. In a political science model of voting behavior in the United States in 1992, we might examine the effect of education on the odds of voting for Clinton versus voting for others. Rather than being additive, the effects are now multiplicative. The idea of interpreting the marginal effect on a transformed η will be discussed in more detail in the next chapter on binary-outcome models.

4. Predicted or Forecast Probabilities Given a Set of Values in the Explanatory Variables. In a classical regression analysis, a researcher may want to compute predicted mean values of Y conditional upon a set of values for the x variables. Similarly, we may want to calculate predicted probabilities from probability models such as the logit and probit, conditional upon a set of values for the x variables. Predicted probabilities are intuitively appealing because these probabilities give an idea of how likely certain types of individuals are to take a certain course of action (or have certain attitudes). For instance, in a multinomial model of contraceptive choice in the United States, we might say that the predicted probability for a white female high school graduate aged 20 to use the pill in the survey year(s) is about .40 (or 40%) and her predicted probability to have sterilization is about .04 (or 4%), whereas the predicted probability for a white female high school graduate aged 30 to use the pill is about .10 (or 10%) and her predicted probability to have sterilization is about .25 (or 25%).

In the generalized linear models framework, predicted probabilities are often derived by calculating the values of μ, using certain values of the x variables. It then becomes a matter of expressing μ, which represents probability in many generalized linear models, as a function of the x variables, because it is specified by the link function of each probability model. There are exceptions. In the Poisson regression model, for instance, μ is not directly related to the probability of an event but to the expectation of event count. This suggests that with the Poisson regression model predicted probabilities are computed differently and that the interpretation in terms of μ gives the expected event count. Therefore, we must calculate the associated predicted probabilities based on the specific link function for our model.

5. Marginal Effect on the Probability of an Event. This last way of interpretation combines features of the previous two ways—marginal effect on η or transformed η and predicted probabilities given a set of values in the explanatory variables. We can interpret the marginal effect of x_k, as expressed by the estimated β_k on the probability of an event, rather than on η or transformed η. This way of interpretation can also be quite useful. Using the contraceptive use example, we might calculate a marginal effect that would suggest that as a white female high school graduate respondent ages one year, the probability of using the pill would decrease by a certain amount.

As with calculating predicted probabilities, we need to determine the level of each of the x variables to compute the marginal effect. In addition, the link function determines the specific formula for calculating the marginal effect on the probability in a particular probability model. The operation will become apparent in later chapters.

These are the five general ways of interpreting parameter estimates from probability models from the perspective of the generalized linear model system. As stated earlier, the first interpretation method is so straightforward that it needs no detailing in later chapters. The second will be discussed only in Chapter 8. Furthermore, variations of these interpretation methods exist. Econometricians are fond of interpretations of elasticity, which is a type of marginal effect. Once the reader has grasped how to interpret, say, parameter estimates as marginal effects on probabilities, the extension to specific interpretations should be a natural one, as in the case of classical linear models when using elasticity. Now let us look at how to interpret binary logit and probit models.

3. BINARY LOGIT AND PROBIT MODELS

As the simplest probability model, binary logit and probit models have only two categories in the response variable—event A or non-A. Such models find many applications in the social sciences. Researchers may want to model the probability of first marriage, dropping out of school, pregnancy, voting for a political candidate, recidivism, occurrence of heart disease, labor force participation, and other events and behavioral patterns. The occurrence and nonoccurrence of these events are the two categories in the dependent variable.

The generalized linear model approach discussed in Chapter 2 can be used to describe models of binary outcomes. In fact, this approach is consistent with an econometric approach (Goldberger, 1964; Maddala, 1983) that assumes an underlying response variable y^* defined by the regression relationship

$$y^* = \sum_{k=1}^{K} \beta_k x_k + \varepsilon .$$

[3.1]

In practice, y^* is unobserved, and ε is symmetrically distributed with zero mean and has its cumulative distribution function (CDF) defined as $F(\varepsilon)$. What we do observe is a dummy variable y, a realization of a binomial process, defined by

$$y = \begin{cases} 1 & \text{if } y^* > 0, \\ 0 & \text{otherwise.} \end{cases}$$

[3.2]

In the model formulated by Equation 3.1, the summed term of βs and xs is not $E(y \mid x_1, \ldots, x_K)$ as in the linear case, but $E(y^* \mid x_1, \ldots, x_K)$.

From these relations we have

$$\text{Prob}(y=1) = \text{Prob}\left(\sum_{k=1}^{K} \beta_k x_k + \varepsilon > 0\right)$$

$$= \text{Prob}\left(\varepsilon > -\sum_{k=1}^{K} \beta_k x_k\right)$$

$$= 1 - F\left(-\sum_{k=1}^{K} \beta_k x_k\right), \qquad\qquad [3.3]$$

where F is a CDF of ε. We may see η, the generalized linear predictor, as the systematic component in y^*, and ε as the random component in y^*. The functional form of F depends on the distribution of, or rather, the assumption made about the distribution of, ε in Equation 3.1. Obviously, the distribution of ε (and our understanding of the distribution) determines the link function of a generalized linear model, another way to represent F.

Logit Models

As mentioned in Chapter 1, many good monographs involving the introduction and treatment of logit models exist. Here I emphasize only the two entries in the Quantitative Applications in the Social Sciences (QASS) series that specifically deal with logit models: Aldrich and Nelson (1984) and DeMaris (1992). Both are good introductions to logit models for both novices and experienced practitioners.

When we assume the random component of the response in the data follows a binomial distribution, we may further assume the logistic distribution for ε. Thus, logit models can apply to the data, and the link function becomes the logit:

$$\eta = \log[\mu/(1-\mu)] \ .$$

Applying this link function to Equation 3.2, we specify a logit model that takes a binary outcome variable. The logit model usually takes two forms. It may be expressed in terms of logit; it may be expressed in terms of event probability. When expressed in logit form, the model is specified as

$$\log\left[\frac{P(y=1)}{1 - P(y=1)}\right] = \sum_{k=1}^{K} \beta_k x_k \ . \qquad\qquad [3.4]$$

Because now we model the probability that event A occurs, or Prob($y = 1$), μ becomes the expected probability that y equals 1. Using Equation 3.3, Equation 3.4 can be transformed into a specification of the logit model of event probability by replacing the general CDF, F, with a specific CDF, L, representing the logistic distribution:

$$\text{Prob}(y=1) = 1 - L\left(-\sum_{k=1}^{K} \beta_k x_k\right) = L\left(\sum_{k=1}^{K} \beta_k x_k\right) = \frac{e^{\sum_{k=1}^{K} \beta_k x_k}}{1 + e^{\sum_{k=1}^{K} \beta_k x_k}} . \qquad [3.5]$$

This represents the probability of an event occurring. For a nonevent, the probability is just 1 minus the event probability or:

$$\text{Prob}(y=0) = L\left(-\sum_{k=1}^{K} \beta_k x_k\right) = \frac{e^{-\sum_{k=1}^{K} \beta_k x_k}}{1 + e^{-\sum_{k=1}^{K} \beta_k x_k}} = \frac{1}{1 + e^{\sum_{k=1}^{K} \beta_k x_k}} . \qquad [3.6]$$

The two forms of logit models are responsible for the different names given to this type of modeling. The model expressed as (3.4) leads to "logit models" because of the logit term, while that expressed as (3.5) leads to "logistic regression" because of the cumulative logistic distribution function. Sometimes in the literature the distinction between the two names of *logit models* and *logistic regression* is based on whether continuous explanatory variables are included in the set of x variables. Some researchers call models with categorical x variables logit models and models with mixed categorical and continuous x variables logistic regression models. Others make no such distinction. Following the generalized linear model tradition, I use "logit models" as the specification of both forms (3.4 and 3.5) regardless of the type of explanatory variables included.

Interpretation of Logit Models

To interpret results from a logit model meaningfully, the model itself must first fit the data. Put differently, the explanatory variables included in the model must be able to explain the response variable significantly better than the model with the intercept only. This is true for all generalized linear models. In a classical regression model, an F test is used; in a logit model (and other probability models), the most commonly used test is the likelihood ratio statistic, which approximately follows the chi-squared distribution (see Aldrich & Nelson, 1984; Greene, 1990; McCullagh & Nelder, 1989; among others). If the model χ^2 as represented by the likelihood ratio statistic indicates that the model fits the data significantly better than the model with the intercept only, we then move on to interpret parameter estimates.

1. Marginal Effect on η or Transformed η. Bypassing the simple interpretation of the sign and significance level of coefficients and predicted values of η or transformed η, we go directly to the arguably easiest and most useful way of interpreting logit models, that of odds and odds ratios. Because linear, additive effects of parameters on the logit are not so intuitively appealing, we exponentiate (take antilogarithm where the base is e) both sides of the link function (Equation 3.4) and have

$$\frac{\text{Prob}(y=1)}{1-\text{Prob}(y=1)} = e^{\eta} = e^{\sum_{k=1}^{K}\beta_k x_k} = \prod_{k=1}^{K} e^{\beta_k x_k}. \qquad [3.7]$$

Now the left-hand side is an odds, and the right-hand side gives the marginal effect of x_k on the odds indicated by $\exp(\beta_k)$. Here the concept of the odds and of the odds ratio, which is the ratio of two odds, becomes central.

Let us use an example for illustrative purposes. The data in Table 3.1 are taken from the National Survey of Children (Morgan & Teachman, 1988, Table 3). For a description of sample, measurement, and substantive issues the reader is referred to Furstenberg, Morgan, Moore, and Peterson (1987). The event of interest here is whether an adolescent (15 and 16 years old) reported ever having had sexual intercourse (yes/no) by the time of the survey. The odds of having had intercourse is $O_{wm} = 43/134 = 0.321$ for white males, $O_{wf} = 26/149 = 0.174$ for white females, $O_{bm} = 29/23 = 1.261$ for black males, and $O_{bf} = 22/36 = 0.611$ for black females. These odds are quite informative. They mean that there are 321 white males who have had intercourse for each one thousand who have not, 174 white females who have had intercourse for each one thousand who have not, 1,261 black males who have had intercourse for each one thousand who have not, and

14

TABLE 3.1
Adolescents Having Ever Had Sexual Intercourse
by Sex and Race; From the National Survey of Children

| Race | Sex | Intercourse | | Odds | Probability | Z of Prob |
		Yes	No			
White	Male	43	134	0.321	0.243	−0.696
	Female	26	149	0.174	0.149	−1.041
	Subtotal	69	283	0.244	0.196	−0.856
Black	Male	29	23	1.261	0.558	0.146
	Female	22	36	0.611	0.379	−0.308
	Subtotal	51	59	0.864	0.464	−0.090

SOURCE: Morgan and Teachman (1988, Table 3). Copyrighted 1988 by the National Council on Family Relations, 3989 Central Ave. NE, Suite 550, Minneapolis, MN 55421. Reprinted by permission.

611 black females who have had intercourse for each one thousand who have not. Also, they indicate that among the same birth cohorts, white females have the smallest likelihood of having had intercourse, black males have the greatest likelihood, black females have the second greatest likelihood, and white males have a likelihood greater than white females and less than black females.

The observed odds suggest that both sex and race are possible factors influencing chances of having had intercourse. We may further construct odds ratios out of odds to make comparisons. In the example, whites regardless of sex have an odds of intercourse less than odds for blacks regardless of sex, $O_w = 69/283 = 0.244$ and $O_b = 51/59 = 0.864$. The odds ratio measures the change in these odds about race. Thus the odds for whites to have had intercourse are 0.282 times as high as the odds for blacks to have had intercourse, because $O_w/O_b = 0.244/0.864 = 0.282$. Conversely, the odds for blacks to have had intercourse are 3.541 times higher than whites, because $O_b/O_w = 0.864/0.244 = 3.541$. (For an examination of the odds ratio comparing males and females, see Morgan & Teachman, 1988.) Similarly, we may construct observed odds ratios comparing the four race/sex-specific groups of white males, white females, black males, and black females by using the odds given in Table 3.1.

The odds ratio has four major desirable properties as a measure of association (Fienberg, 1980; Morgan & Teachman, 1988):

a. It has a clear interpretation such that the move from one odds to the next induces a multiplicative change as indicated by the odds ratio. An odds

TABLE 3.2
Logit Model Estimated for Data in Table 3.1

Variable	$\hat{\beta}$	$se(\hat{\beta})$	p	$exp(\hat{\beta})$
White	−1.314	0.226	0.000	0.269
Female	−0.648	0.225	0.004	0.523
Constant	0.192	0.226	0.365	1.212
LR Statistic	37.459			
df	2			

ratio greater than 1 indicates an increased chance of an event occurring versus not (e.g., the odds for blacks to have had intercourse are 3.541 times higher than for whites), and an odds ratio less than 1 indicates a decreased chance of an event occurring versus not occurring (e.g., the odds for whites to have had intercourse are 0.282 times as high as for blacks). Recall that 0 instead of 1 is the threshold between negative and positive effects for additive models such as classical regression, and that if we take the logarithm of an odds ratio, it becomes log-odds ratio, which has the same properties of additive effects including the sign of effect being demarcated by 0.

b. It is invariant with respect to the ordering of the variables except for its "sign." For instance, $O_b/O_w = 1/(O_w/O_b)$ if we switch the order of the variables in the comparison.

c. It is invariant with respect to frequency multiplications in the variables. If we increase sample size in the intercourse example, we should get approximately the same odds and hence the odds ratios.

d. It can be used in studying polytomous response variables and models with multiple explanatory variables. This last property generalizes from the same property of odds ratios in multidimensional contingency tables.

These properties facilitate the interpretation of parameter estimates from logit models as well. Table 3.2 gives the coefficient estimates from a logit model applied to the data in Table 3.1. First, the likelihood ratio statistic shows that the model is significantly different from the null or intercept-only (or know-nothing) model by a χ^2 test (37.459 with 2 degrees of freedom). Both variables have an estimate significantly different from 0, as judged by the size of $\hat{\beta}$ relative to its estimated asymptotic standard error, and further indicated by the column labeled p, which gives the upper bound of the probability of making a Type I error. (Such a significance test is usually performed by assessing a test statistic—either Wald χ^2 or Z statistic, depending on the statistical software package used. I used PROC LOGIS-TIC in SAS 6.08, which gives Wald χ^2 statistic.)

Exponentiating $\hat{\beta}$ gives the last column labeled "exp($\hat{\beta}$)" in Table 3.2. The values in this column give the expected change in the transformed η, or the odds of having an event occurring versus not occurring, per unit change in an explanatory variable, other things being equal. The same interpretation applies to both dummy and continuous variables. We will focus on dummy variables in the present example and examine an example of continuous variables in the next chapter. In the current example, a unit change in either the race or the sex variable (from 0 to 1) really indicates switching from black to white in the dummy variable called "white" or from male to female in the dummy variable called "female." Therefore, interpreting the marginal effect on the odds of an event occurring is tantamount to examining the odds ratio between the odds of each event occurring.

Using this interpretation, the odds for whites to have had intercourse are estimated to be 0.269 times as high as blacks, other things being equal. This estimate is somewhat lower than the observed odds ratio of 0.282 calculated earlier. If the interaction between race and sex is included in the model, the difference between the observed and the estimated odds ratio of race within each category of sex (or of sex within each category of race) will disappear.

If a researcher is interested in comparing blacks to whites instead of vice versa, the estimate can be easily derived from current estimates without refitting the model. This follows from properties (a) and (b) of the odds ratio. All we need to do is to change the sign of the relevant $\hat{\beta}$, that is, changing -1.314 to 1.314 and exponentiating it, yielding an odds ratio of 3.721. We may apply the same operation to the "female" estimate and get 1.912. Thus, the odds for males to have had intercourse are 1.912 times higher than females, other things being equal. We can see that interpreting the marginal effect of an x variable on the odds of an event occurring is an easy and flexible way of interpretation.

2. Predicted Probabilities Given a Set of Values in the Explanatory Variables. Because of the ease in understanding the probability of an event occurring, it is also very appealing to compute the predicted probability for a given set of values of the x variables. A primary issue is what values to choose for x variables. If we have many independent variables, some of which are categorical, some continuous, it will probably be advisable to focus on one or two variables of interest at a time and set the values in other variables at their sample means. If one of the two variables we have selected is discrete and the other is continuous, we may choose to plot the predicted probability values against the continuous x variable within each category of the discrete variable. We will see an example of this in Chapter 4 when sequential logit models are examined. How many data points do we use in the continuous variable to make predictions? That is chiefly up to one's judgment. The issue boils down to a trade-off between simplicity and clarity on the one hand and specificity and accuracy on the other. In the current example of sexual intercourse, we might choose to calculate predicted probabilities at all values of the x variables because we have only two dichotomous variables in the model.

Second, how do we calculate predicted probabilities? We use Equation 3.5, which expresses event probability as a function of βs and xs. We will have a predicted probability for each subgroup of white females, white males, black females, and black males. Writing out the equation for predicting the probability of having had intercourse among white female adolescents, we have

$$\text{Prob}_{wf}(y = 1) = \frac{e^{0.192 \cdot \text{constant} - 1.314 \cdot \text{white} - 0.648 \cdot \text{female}}}{1 + e^{0.192 \cdot \text{constant} - 1.314 \cdot \text{white} - 0.648 \cdot \text{female}}}$$

$$= \frac{e^{0.192 \cdot 1 - 1.314 \cdot 1 - 0.648 \cdot 1}}{1 + e^{0.192 \cdot 1 - 1.314 \cdot 1 - 0.648 \cdot 1}} = 0.146, \qquad [3.8]$$

where the first 1 in the numerator is the value of the constant, the second is the code for white in the white dummy variable, and the third is the code for female in the female dummy variable, and 0.192, −1.314, and −0.648 are the coefficient estimates for the intercept, the race variable, and the sex variable, respectively. Similarly, the predicted probability of having had intercourse for white male adolescents is

$$\text{Prob}_{wm}(y = 1) = \frac{e^{0.192 \cdot 1 - 1.314 \cdot 1 - 0.648 \cdot 0}}{1 + e^{0.192 \cdot 1 - 1.314 \cdot 1 - 0.648 \cdot 0}} = 0.246. \qquad [3.9]$$

The predicted probability of having had intercourse for black females is

$$\text{Prob}_{bf}(y = 1) = \frac{e^{0.192 \cdot 1 - 1.314 \cdot 0 - 0.648 \cdot 1}}{1 + e^{0.192 \cdot 1 - 1.314 \cdot 0 - 0.648 \cdot 1}} = 0.388. \qquad [3.10]$$

Finally, the predicted probability of having had intercourse among black males is

$$\text{Prob}_{bm}(y = 1) = \frac{e^{0.192 \cdot 1 - 1.314 \cdot 0 - 0.648 \cdot 0}}{1 + e^{0.192 \cdot 1 - 1.314 \cdot 0 - 0.648 \cdot 0}} = 0.548 . \qquad [3.11]$$

These predicted probabilities tell us proportionally how many members of each group may have had intercourse, thereby giving an easy, intuitive understanding. Based on the logit model, while about 55% of black male adolescents are predicted to have had intercourse, only about 15% of white female adolescents are predicted to have had the experience. The reader may want to compare the predicted probabilities for the subgroups to those observed probabilities given in Table 3.1. The reader may also want to compare the ordering of the predicted probabilities to that of the observed odds given earlier.

3. *Marginal Effect on the Probability of an Event.* Instead of examining the marginal effect of an x variable on the odds, we may also look at the marginal effect of the variable on the probability of the event. The effect is given by the following equation:

$$\frac{\partial \text{Prob}(y = 1)}{\partial x_k} = \frac{e^{\sum_{k=1}^{K} \beta_k x_k}}{\left(1 + e^{\sum_{k=1}^{K} \beta_k x_k}\right)^2} \beta_k = P(1 - P)\beta_k , \qquad [3.12]$$

where the round-backed d, ∂, indicates partial derivative or marginal effect, P is short for Prob($y = 1$), and $1 - P$ represents Prob($y = 0$), as given in Equations 3.5 and 3.6. Unlike the interpretation of the effect on the odds, however, which is invariant to the values of independent variables, the marginal effect on probability changes with the values of the xs and hence with the probability associated with the values of the xs (see also DeMaris, 1990; Greene, 1990; Hanushek & Jackson, 1977). In addition, partial derivatives give only a rough approximation of the marginal effect of a dummy variable on probability, although they give a close approximation of the marginal effect of a continuous variable on probability. For detailed explanations of why the (partial) derivative is an approximation of the marginal effect of a continuous variable, I refer the reader to standard calculus texts such as the one by Hoffman and Bradley (1989). The derivative is commonly used for calculating the effect on y given a unit change in x because it is easier than computing the probability before and after the unit change. The use of derivatives has produced some confusion and thus debate in the social sciences (e.g., DeMaris, 1993; Roncek, 1991, 1993).

The derivative of a function is defined for continuous variables only. Thus, derivatives are not defined in principle for discrete variables such as a dummy variable with 0 and 1 values only, although they can give a rough approximation of the marginal effect of a dummy variable in practice. A better estimate of such marginal effect is computed by taking the difference of the predicted probability conditional on each of the two categories in the dummy variable (see Greene, 1990; Roncek, 1991). We will use the dummy variables in the simple example for illustrating the general method of interpreting the marginal effect on probability, and compare the marginal effects thus gotten with those estimated from taking the difference in predicted probabilities. Interested readers who want to find out more about the difference between dummy and continuous variables in interpreting the marginal effect on response probability are referred to the relevant section in Chapter 5.

For the current example, we utilize the predicted probabilities for the four subgroups (which are at all the possible values of the x variables in the model) calculated previously to arrive at various marginal effects. Specifically, the difference in the probability of having had intercourse between white females and black females is estimated to be -0.242 or $0.146 - 0.388 = -0.242$. If we calculate the partial derivative using the probability of black females, the marginal effect of race is about -0.312 or $P(1 - P)\beta_k = 0.388(1 - 0.388)(-1.314) = -0.312$. This estimate is about 0.07 off the actual difference between the predicted probabilities. Similarly, the difference in the probability of having had intercourse between white females and white males is estimated to be -0.100 or $0.146 - 0.246 = -0.100$. If we calculate the partial derivative using the probability of white males, the marginal effect of sex is about -0.120 or $0.246(1 - 0.246)(-0.648) = -0.120$. This estimate is about 0.02 off the actual difference between the corresponding probabilities. Furthermore, the difference in the probability of having had intercourse between white males and black males is estimated to be -0.302 or $0.246 - 0.548 = -0.302$. If we calculate the partial derivative using the probability of white males, the marginal effect of race is about -0.325 or $0.548(1 - 0.548)(-1.314) = -0.325$. This estimate is about 0.023 off the actual difference between the corresponding probabilities. Similarly, the difference in the probability of having had intercourse between black females and black males is estimated to be -0.160

or $0.388 - 0.548 = -0.160$. If we calculate the partial derivative using the probability of black females, the marginal effect of sex is about -0.161 or $0.548(1 - 0.548)(-0.648) = -0.161$. This estimate is about 0.001 off the actual difference between the corresponding probabilities. For a continuous variable, we will make an interpretive statement such as given a unit change (or increase) in x_k, other things being equal, the event probability is expected to change by an amount $P(1 - P)\beta_k$ approximately. In sum, the marginal effect of an explanatory variable on probability varies with the level of the event probability, which in turn depends on the values in the set of explanatory variables.

The discrepancy between the two methods—taking partial derivatives and the difference between two predicted probabilities—has to do with taking partial derivatives of dummy variables, because values are either 0 or 1, whereas a partial derivative calculates an expected instantaneous change in probability given a change in x_k so small that it approaches zero. In a continuous variable, a unit change approximates a small change, thereby the partial derivative approximates the marginal effect; while in a dummy variable the only possible change is from 0 to 1 or from 1 to 0, a 100% change. Therefore, the issue becomes clear that how good the approximation is depends on how fine the measurement is, even in the case of a continuous variable. Although interpreting the effect of a dummy variable by using partial derivatives does not present any problem in linear models whose link function is the identity, discrepancies arise for other generalized linear models because their link functions are nonlinear. Thus taking partial derivatives with respect to a dummy variable tends to overestimate the marginal effect, and care needs to be taken when interpreting partial derivatives of dummy variables. The values may differ by a noticeable, though not large, amount (as in the current example). In Chapter 5 a comparison of using partial derivatives and the difference in predicted probabilities will be made for both dummy and continuous variables. The case becomes transparent once a continuous variable is brought into the comparison.

An alternative interpretation, however, is to ignore the dependency on the values in xs and calculate marginal effects at hypothetical levels of event probabilities. We may ask, what is the marginal effect for individuals with event probability identical to observed white females and black females probabilities (or proportions), respectively? Or more generally, we can calculate the marginal effect at p values equal to 0.00, 0.25, 0.50, 0.75, and 1.00, respectively. Doing so gives the range of marginal effects but loses sight of reality, because only a certain range of probabilities is possible, given the sample, and these probabilities are linked to individuals with certain characteristics. Overall, with the ease of the first two ways of interpreting logit model results, interpreting marginal effects on probability is less common in empirical social science research.

Probit Models

The probit model represents another type of widely used statistical model for studying data with binomial distributions. Its employment in the social sciences goes back at least to econometrics in the early 1960s (see Goldberger, 1964). For good introductory treatment, see Aldrich and Nelson (1984), Greene (1990), and Maddala (1983). Probit models are generalized linear models with a probit link:

$$\eta = \Phi^{-1}(\mu).$$

The inverse of the normal CDF is in effect a standardized variable, or a Z score. As with the logit model, the probit model is used for studying a binary outcome variable. We may express probit models in probability,

$$\text{Prob}(y = 1) = 1 - F\left(-\sum_{k-1}^{K} \beta_k x_k\right) = F\left(\sum_{k=1}^{K} \beta_k x_k\right) = \Phi\left(\sum_{k=1}^{K} \beta_k x_k\right), \quad [3.13]$$

where the more general form of cumulative distribution function, F, is replaced by the standard normal cumulative distribution function, Φ. Unlike the logit model, which may take on two major forms—one expressing the model in logit (and a transformed version expressed in odds) and the other expressing the model in event probability—the probit model takes on only one intuitively meaningful form, because a probit model expressed in η is a linear regression of the Z score of the event probability. The equation for probability of nonevent is then

$$\text{Prob}(y = 0) = 1 - \Phi\left(\sum_{k=1}^{K} \beta_k x_k\right). \quad [3.14]$$

The equation can be readily derived from Equation 3.13, because the response is a binary outcome.

TABLE 3.3

Probit Model Estimated for Data in Table 3.1

Variable	$\hat{\beta}$	$se(\hat{\beta})$	p
White	−0.789	0.144	0.000
Female	−0.377	0.131	0.004
Constant	0.106	0.138	0.584
LR Statistic	37.379		
df	2		

Interpretation of Probit Models

The probit model's lack of alternative forms limits the flexibility in its interpretation. This point is shown by the interpretation of the effect x_k on η.

1. Marginal Effect on η. Because there is no easy way to transform η, which takes the form of the inverse of the standard normal cumulative distribution function in the probit model, interpreting the marginal effect on η means interpreting linear, additive effects of the x variables on the inverse of the standard normal cumulative distribution function. Even though we may see the inverse of the standard normal CDF as a Z score and the marginal effect on η as an effect on the Z score, such interpretation is still far from being intuitively obvious. In Table 3.3, I present results from a probit model applied to the data in Table 3.1. The effect for race variable called white (−0.789), for instance, is the difference between the expected Z score for whites and that for blacks. This expected difference in Z score should be close to the observed difference. Taking the difference between the observed Z score for blacks (both sexes) and that for whites (both sexes) gives −0.766, because −0.856 − (−0.090) = −0.766, based on the values in the last column in Table 3.1. Thus the expected effect on the Z score is different from the observed value by 0.023 only.

We might as well go one step further by transforming Z scores into probabilities and interpreting the marginal effects on event probabilities. This leaves us with two useful ways of interpretation—predicted probabilities and the marginal effects on event probabilities. Because interpreting the parameter estimates directly (as compared with interpreting the effect on the transformed η or the odds in logit models) by noting their effects on Z scores lacks intuitive appeal, I save most of the discussion of the results for the remaining two ways of interpretation.

2. Predicted Probabilities Given a Set of Values in the Explanatory Variables. The interpretation of probit results through predicted probabilities is identical to that of logit results except for the difference in their cumulative distribution functions. Equations 3.13 and 3.14 give event and nonevent probabilities. The predicted probabilities for the four subgroups are calculated as follows:

$$\text{Prob}_{wf}(y = 1) = \Phi(0.106 \cdot \text{constant} - 0.789 \cdot \text{white} - 0.377 \cdot \text{female})$$

$$= \Phi(0.106 \cdot 1 - 0.789 \cdot 1 - 0.377 \cdot 1) = 0.145, \qquad [3.15]$$

$$\text{Prob}_{wm}(y = 1) = \Phi(0.106 \cdot 1 - 0.789 \cdot 1 - 0.377 \cdot 0) = 0.247, \qquad [3.16]$$

$$\text{Prob}_{bf}(y = 1) = \Phi(0.106 \cdot 1 - 0.789 \cdot 0 - 0.377 \cdot 1) = 0.393, \qquad [3.17]$$

$$\text{Prob}_{bm}(y = 1) = \Phi(0.106 \cdot 1 - 0.789 \cdot 0 - 0.377 \cdot 0) = 0.542. \qquad [3.18]$$

Unlike the calculation of predicted probabilities in logit models, which can be easily done with a hand-held calculator, the involvement with Φ means that a statistical software package is to be used. Fortunately, most widely used packages such as SAS, SPSSX, BMDP, and LIMDEP can perform the calculation of such functions easily. For a few quick calculations, one may also look up the statistical table for area under the normal curve (relating Z scores and probability).

The predicted probabilities generated from the probit model are almost identical to those from the logit model, and they differ by less than 5 per thousand. Thus the results will lead to similar conclusions. If the calculation of Φ function is easily accessible, these probabilities can be as easy to compute as those using the logit model.

3. Marginal Effect on the Probability of an Event. Similar to the third way of interpretation in the logit model, here we examine partial derivatives of probability with respect to an independent variable, x_k (see Greene, 1990; Maddala, 1983):

$$\frac{\partial \text{Prob}(y = 1)}{\partial x_k} = \phi\left(\sum_{k=1}^{K} \beta_k x_k\right)\beta_k, \qquad [3.19]$$

where ϕ indicates the standard normal probability density function. Applying Equation 3.19 to the parameter estimates in Table 3.3, we obtain the following results:

Because the result of $\phi(\cdot)$ is a function of all the xs, we can only compute marginal effects by assigning certain values to the xs. Again, we may calculate marginal effects for the four subgroups (which include all possible values of the x variables in the example). Let us interpret the effects of race and sex in a way similar to interpreting the effect of a continuous variable, with the understanding that these will be rough approximations only. For black females [with $\phi(0.106 \cdot 1 - 0.789 \cdot 0 - 0.377 \cdot 1) = 0.385$], the marginal effect of race is $(0.385)(-0.789) = -0.303$, suggesting that for a black female the unit change in race (were she a white female) would induce a decrease in event probability of about 0.303, holding sex constant. Similarly, for white males [with $\phi(0.106 \cdot 1 - 0.789 \cdot 1 - 0.377 \cdot 0) = 0.316$], the marginal effect of sex is $(0.316)(-0.377) = -0.119$, suggesting that for a white male the unit change in sex (were he a white female) would induce a decrease in event probability of about 0.119, holding race constant. For black males [with $\phi(0.106 \cdot 1 - 0.789 \cdot 0 - 0.377 \cdot 0) = 0.397$], the marginal effects of race and of sex are -0.313 and -0.150, respectively, suggesting that for a black male the unit change in race (were he a white male) would induce a decrease in event probability of about 0.313, holding sex constant, and the unit change in sex (were he a black female) would induce a decrease in event probability of about 0.150, holding race constant. Again, the partial derivative of dummy variables overestimates marginal effect, if compared with the corresponding differences between predicted probabilities given by Equations 3.15 through 3.18, as in the logit example. The reader may want to calculate the more correct and accurate marginal effects by taking the difference in probability to compare results. Comparing the marginal effects based on Equation 3.19 or the difference in probability with the corresponding effects from the logit model, the differences are present but small, suggesting either method will lead to similar substantive conclusions.

Logit or Probit Models?

Given the similarities between the two types of models, either model will give identical substantive conclusions in most applications. In fact, one can go from one set of estimates to the other. If one multiplies a probit estimate by a factor, one gets an approximate value of the corresponding logit estimate. This factor is normally believed to be $\pi/\sqrt{3} = 1.814$ (see e.g., Aldrich & Nelson, 1984). However, Amemiya (1981) proposed, by trial and error, a value of 1.6 that should approximate more closely. The most accurate value of the factor lies somewhere in the neighborhood of these two values. There are situations, however, where estimates from logit and probit models may differ substantially, and in such cases care must be taken in choosing the more appropriate model. These are cases with an extremely large number of observations and with a heavy concentration of observations in the tails of the distribution (Amemiya, 1981). Logit models are more appropriate for distributions with heavier tails.

4. SEQUENTIAL LOGIT AND PROBIT MODELS

Sequential logit and probit models, also called sequential-response, hierarcl.ical-response, or nested logit and probit models, are natural extensions of the binary logit and probit models. What applies to binary logit and probit models applies to sequential models as well, because a sequential model most often is really a sequence of binary-outcome models, be they logit or probit. Exceptions do exist in which a stage in the sequence is another probability model such as a multinomial-response model. Due to the similarities between logit and probit models and the popularity and greater flexibility in interpretation of logit models, we focus on sequential logit models for empirical examples.

The Model

Sometimes the outcomes in a response variable are polytomous, but they are not a collection of purely discrete, unrelated categories. Rather, the response categories can be perceived as a sequence with stages. The response in a later stage is nested in the response in an earlier stage. For example, the decision to get married can be perceived as a sequence with two stages—whether someone plans to ever marry and, if yes, whether the marriage will be before the completion of a certain amount of education such as finishing high school or college.

Maddala (1983) considers two examples of sequential-response models in the social science literature. One involves a sequence of educational attainment:

$y = 1$ if the individual has not finished high school,

$y = 2$ if the individual has finished high school but not college,

$y = 3$ if the individual has completed college but not a professional degree,

$y = 4$ if the individual has a professional degree.

The related probabilities can be written as (see Amemiya, 1975; Maddala, 1983):

$$P_1 = F\left(\sum_{k1}^{K1} \beta_{k1} x_{k1}\right),$$

$$P_2 = \left[1 - F\left(\sum_{k1}^{K1} \beta_{k1} x_{k1}\right)\right] F\left(\sum_{k2}^{K2} \beta_{k2} x_{k2}\right),$$

[4.1]

$$P_3 = \left[1 - F\left(\sum_{k1}^{K1} \beta_{k1} x_{k1}\right)\right]\left[1 - F\left(\sum_{k2}^{K2} \beta_{k2} x_{k2}\right)\right] F\left(\sum_{k3}^{K3} \beta_{k3} x_{k3}\right),$$

$$P_4 = \left[1 - F\left(\sum_{k1}^{K1} \beta_{k1} x_{k1}\right)\right]\left[1 - F\left(\sum_{k2}^{K2} \beta_{k2} x_{k2}\right)\right]\left[1 - F\left(\sum_{k3}^{K3} \beta_{k3} x_{k3}\right)\right],$$

where the $k1$, $k2$, $k3$, and $k4$ subscripts indicate the sets of x variables included in Stage 1, 2, 3, and 4, respectively. The parameters β_{k1} can be estimated by dividing the entire sample into two groups—those who have not finished high school and finished high school. The Stage 2 parameter β_{k2} can be estimated from the subsample of high school graduates by dividing it into two groups—those who have not finished college and finished college. The Stage 3 parameter β_{k3} can be estimated from the college graduates by dividing them into two groups—those who have no professional degree and have professional degree. Notice that as with the binary model, we will always need to estimate $J - 1$ number of sets of parameters, with J being the total number of categories in the response. In each stage of the sequential model, a binary model can be estimated by the logit or probit method.

Another example is provided by McCullagh and Nelder (1989) with an epidemiological study of mortality due to radiation. The study has three stages: Individuals exposed and not-exposed to radiation are classified at the end of the study as alive or dead in Stage 1. In Stage 2, deaths are further studied by classifying into deaths due to cancer and other causes of death. In Stage 3, deaths due to cancer are further classified into Leukemia deaths and deaths due to other cancers. The model can be specified identically to the one of educational attainment, with the same number of sets of β_ks to estimate.

Sometimes the dichotomous outcomes are not nested so orderly in one branch in the sequence of the decision-making tree only. The model on automobile demand studied by Cragg and Uhler (1975) and discussed by Maddala (1983) provides another kind of sequential decision making. The model consists of a series of binary choices:

$y_1 = 1$ if the individual has acquired a car,
$y_1 = 2$ if the individual has not acquired a car;
$y_2 = 1$ if the individual has acquired a car to replace an old car,
$y_2 = 2$ if the individual has acquired a car to add to an old car;
$y_3 = 1$ if the individual has not acquired a car but sold an old car,
$y_3 = 2$ if the individual has neither acquired a car nor sold an old car.

There are four probabilities of interest:

P_1 = probability of replacing a car,
P_2 = probability of adding a new car,
P_3 = probability of selling a car,
P_4 = probability of no change.

The probabilities can be defined with β_{k1}, β_{k2}, and β_{k3} that are related to the three ys. The parameters β_{k1} can be estimated from the whole sample by dividing it into those who have acquired new cars and those who have not. The parameters β_{k2} can be estimated from the subsample of new car purchasers by dividing it into those who have replaced cars and those who have added cars. The parameters β_{k3} can be estimated from the subsample of those who have not acquired a car by dividing it into those who have sold old cars and those who have done nothing. We can write these probabilities as

$$P_1 = F\left(\sum_{k1}^{K1} \beta_{k1} x_{k1}\right) F\left(\sum_{k2}^{K2} \beta_{k2} x_{k2}\right),$$

$$P_2 = F\left(\sum_{k1}^{K1} \beta_{k1}x_{k1}\right)\left[1 - F\left(\sum_{k2}^{K2} \beta_{k2}x_{k2}\right)\right],$$

$$P_3 = \left[1 - F\left(\sum_{k1}^{K1} \beta_{k1}x_{k1}\right)\right]F\left(\sum_{k3}^{K3} \beta_{k3}x_{k3}\right),$$ [4.2]

$$P_4 = \left[1 - F\left(\sum_{k1}^{K1} \beta_{k1}x_{k1}\right)\right]\left[1 - F\left(\sum_{k3}^{K3} \beta_{k3}x_{k3}\right)\right].$$

Again, the sequential model can be estimated using a series of binary logit or probit models.

Sequential logit models have found straightforward applications in social science research. Plotnick (1992) studies premarital pregnancy and its resolution by using a sequential logit model with two stages. In the first stage he uses a binary logit model, and in the second stage he uses a multinomial logit model. (The reader is referred to Chapter 6 for multinomial logit models.)

$y_1 = 1$ if a teenage girl has had a premarital pregnancy,

$y_1 = 2$ if a teenage girl has not had a premarital pregnancy;

$y_2 = 1$ if a teenage girl has had a premarital pregnancy and an abortion,

$y_2 = 2$ if a teenage girl has had a premarital pregnancy and married before the birth,

$y_2 = 3$ if a teenage girl has had a premarital pregnancy and a premarital birth.

The probabilities of interest in this sequential model can be expressed by a single equation:

$$P_{ij} = P_i \cdot P_{j|i},$$ [4.3]

where P_i refer to the probabilities for the outcomes of y_1, and $P_{j|i}$ refer to the conditional probabilities for the outcomes of y_2, and P_{ij} are final probabilities.

In studying spouse's functioning in personal social networks, Liao and Stevens (forthcoming) model married individuals' likelihood of considering their spouse as the first-choice person to talk with about important matters, using a two-stage sequential model:

$y_1 = 1$ if the married individual has included the spouse in the personal social network,

$y_1 = 2$ if the married individual has not included the spouse in the personal social network;

$y_2 = 1$ if the married individual has included the spouse in the network and regards him/her as first choice,

$y_2 = 2$ if the married individual has included the spouse in the network but regards him or her as a later choice.

A sequential logit model with a binary logit model in each stage is used. Probabilities of interest can also be defined similarly to Equation 4.3—with P_i related to y_1 (inclusion of spouse in the network), $P_{j|i}$ related to y_2 (regarding spouse as a first or a later choice), and P_{ij} as the probability of final outcomes.

One important point about sequential models is that the probability of choice at each stage should be independent of the probability of the choice at the other stages (Maddala, 1983). In other words, the y_1, y_2, and so forth, variables should be conceptually distinct and statistically independent from each another. Also, the assumed sequence may be one of the possible models determined by theory. For example, a researcher may assume that a citizen decides whether or not to vote, and then decides which candidate to support, but the opposite sequence is also possible, as is the possibility that the two decisions are made simultaneously or interrelatedly.

Interpretation of Sequential Logit and Probit Models

Because a sequential logit model consists of a series of binary logit models, interpretation of a sequential logit model follows that of a binary logit model. Similarly, interpretation of a sequential probit model follows that of a binary probit model. In the following I mainly highlight the differences.

1. Marginal Effect on η or Transformed η. Each stage in the sequential model has its own η, transformed or not. This is evidenced by the number of β_{k1}, x_{k1}; β_{k2}, x_{k2}; β_{k3}, x_{k3}, and so on, with each set of $\beta_{k\cdot}$ and $x_{k\cdot}$ corresponding to a unique η or transformed η. Therefore, the marginal

TABLE 4.1
Sequential Logit Model Estimated for General Social Survey, 1985

| Response Variable | Spouse Included? (y_1) | | | | Spouse Named First? (y_2) | | | |
| Sample | Married Males | | | | Married Males With $y_1 = 1$ | | | |
x Variable	$\hat{\beta}$	se($\hat{\beta}$)	p	$\bar{x}$	$\hat{\beta}$	se($\hat{\beta}$)	p	$\bar{x}$
Size of Network	0.207	0.074	.005	2.94	0.075	0.140	0.596	3.25
1-Person Network?	—	—	—	—	9.359	14.184	0.509	0.20
Age	0.079	0.057	.171	47.81	0.068	0.098	0.488	45.07
Age Squared	−0.001	0.001	.083	2,517.4	−0.001	0.001	0.418	2,232.5
White?	0.897	0.445	.044	0.92	0.759	0.879	0.388	0.95
Catholic?	0.166	0.267	.533	0.31	−0.247	0.369	0.503	0.32
# of Children	−0.178	0.080	.026	2.45	−0.222	0.134	0.098	2.25
Education	0.046	0.072	.528	12.50	−0.032	0.089	0.723	13.10
Rural Residence?	−0.485	0.259	.061	0.31	0.107	0.390	0.784	0.25
Age at Marriage	−0.012	0.027	.653	23.60	−0.027	0.043	0.539	23.53
Marital Sameness								
Same Religion?	0.818	0.379	.031	0.89	1.374	0.567	0.016	0.91
Educational Diff.	−0.196	0.218	.367	1.83	−0.692	0.381	0.069	1.73
Ed. × Ed. Diff.	0.012	0.017	.472	23.40	0.053	0.026	0.039	24.22
Constant	−2.558	1.653	.122	—	−1.464	2.528	0.562	—
Model χ^2 and df	62.710 & 12				60.913 & 13			
N	370				232			

SOURCE: Liao and Stevens (forthcoming, Tables 1 and 2).
NOTE: Dummy variables are indicated with question marks, and coded 1 if "yes," 0 otherwise.

effect on η or on transformed η is interpreted within each stage typically. The interpretation then becomes identical to that in a binary logit or probit model. We now examine an example of the effect of a continuous variable on the odds.

Let us look at the social network example to illustrate how to use predicted probabilities in a sequential logit model. Table 4.1 presents parameter estimates, their standard errors, and related Type I error probabilities estimated from a sequential logit model, using data from the 1985 General Social Survey. For further details see Liao and Stevens (forthcoming).

To review, the question is the inclusion of one's spouse in the close personal network for discussion of important matters. The decision is conceived as having two steps: first, whether one's spouse is included in this personal social network for discussing important matters; then, if yes, whether one's spouse is considered as the first person one wants to talk with about important matters.

One variable that has a statistically significant effect in Stage 1 is the number of children the respondent has ever had. The effect of the number of children on the odds of including spouse in the discussion network is $\exp(-0.178) = 0.837$. Thus, other things being equal, an additional child among married males would reduce the odds of including spouse in the close social network by a factor of 0.837. In other words, the odds for a married male with one more child to include spouse in his close social network are only about 0.837 times as high as a married male without this additional child. The interpretation is not affected by the nature of sequential response.

2. Predicted Probabilities Given a Set of Values in the Explanatory Variables. Predicted probabilities are calculated in the same way as those in binary logit and probit models. The only difference from binary-outcome models is that predicted probabilities in a sequential model involve a product of probabilities from the relevant stages while those in a binary-outcome model can be directly calculated from the one-stage model. The four examples in the previous section can serve as a guide to various sequential-response models. The key is to write out which decision is conditional upon what (sub)sample by writing the definitions of *y*s out clearly. This done, the equation(s) for probabilities will follow naturally. Drawing a decision-making tree may also help in determining which probabilities should be included in the products of probabilities from each stage.

In the social network example, we focus on the effects of heterogamy—differences in personal backgrounds of the spouses. Three variables measure homogamy or heterogamy: whether the marital partners have the same religious affiliation, the difference in completed education in years, and the interaction between difference in educational attainment of the spouses and the respondent's level of education, also measured in completed years of education. These variables are in the block under the label "Marital Sameness" in Table 4.1.

The predicted unconditional probability for various levels of homogamy or heterogamy is defined as

$$\text{Prob}(y_2 = 1) = \text{Prob}(y_1 = 1) \cdot \text{Prob}(y_2 = 1 \mid y_1 = 1) = \frac{e^{\sum_{k1}^{K1} \beta_{k1} x_{k1}}}{1 + e^{\sum_{k1}^{K1} \beta_{k1} x_{k1}}} \frac{e^{\sum_{k2}^{K2} \beta_{k2} x_{k2}}}{1 + e^{\sum_{k2}^{K2} \beta_{k2} x_{k2}}} \cdot$$

$$[4.4]$$

32

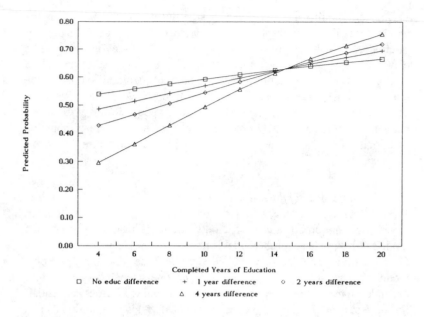

Figure 4.1. Likelihood a Male Whose Spouse Reports the Same Religion Names
Her First in Discussion Network
SOURCE: Table 4.1

The parameter estimates β_{k1} are reported in column 1 and β_{k2} in column 5
of Table 4.1. Because the patterns of inclusion of spouse differ by sex, the
analysis is stratified by sex. Only the results for males are reproduced here.

For all x variables except education and the three related to homogamy,
sample means are used in the calculation. For the dummy variable of
religious homogamy, 1 and 0 are used; for difference in education, four
empirically most common levels are used—no difference, 1 year difference, 2 years difference, and 4 years difference; for respondent's completed
years of education, nine levels are used—4 to 20 years of education
incremented by 2 years. Four and 20 years of completed education are the
observed minimum and maximum values in the sample, and an increment
of 2 should be fine enough to describe any major change in the curve. There
are, in total, $2 \times 4 \times 9 = 72$ predicted probabilities. The rationale for
calculation is the same as that in the binary logit model. Thus no detailed
steps of calculation are reported. Values of predicted probabilities are not
shown, either, since with 72 predicted probabilities, graphs are more intui-

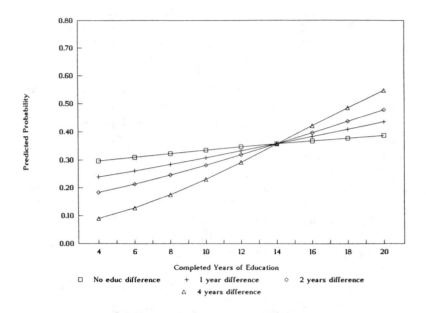

Figure 4.2. Likelihood a Male Whose Spouse Reports a Different Religion Names Her First in Discussion Network
SOURCE: Table 4.1

tively appealing than tables, as shown in Figures 4.1 and 4.2. Because there is an interaction term between education and educational difference, the predicted probability is graphed against the level of education, with each level of educational difference represented by a curve. Because it is confusing to have more than five or six curves in one graph, separate graphs with the same range of scales are made for religious homogamy and heterogamy.

Once the graphs are constructed based on calculated probabilities, researchers can easily interpret the effect of homogamy/heterogamy. Although education has an overall positive effect on the final, unconditional predicted probability of naming his spouse as first choice, this effect increases with the level of difference in education: The difference in predicted probabilities between 0 and 4 years of educational difference can be greater than 0.20 at the same level of respondent's education. Moreover, although the same pattern of interaction between education and difference in education can be observed for both religious homogamy and heterogamy, religiously homogamous males are much more likely to name their spouse as their first choice to discuss important matters with. A comparison of Figures 4.1 and 4.2 reveals that, despite some minor differences in the curves, the group of four curves for religious homogamy is about 0.20 higher in probability than the four for religious heterogamy.

3. Marginal Effect on the Probability of an Event. Interpreting the marginal effect of an independent variable on event probability in a sequential model is a natural extension of the interpretation of results from binary-outcome models, again because the sequential model under examination usually consists of a series of binary-outcome models. Therefore, partial derivatives of the overall probability in the sequence with respect to x_k can be worked out using respective partial derivatives from the binary-response model in each of the stages involved.

Let us use the social network example to illustrate the interpretation. Recall from Equation 3.12 that the marginal effect of a unit change in x_k on event probability in a binary logit model is

$$\frac{\partial \text{Prob}(y=1)}{\partial x_k} = \frac{e^{\sum_{k=1}^{K} \beta_k x_k}}{\left(1 + e^{\sum_{k=1}^{K} \beta_k x_k}\right)^2} \beta_k = P(1-P)\beta_k, \qquad [4.5]$$

where again the round-backed d, ∂, indicates partial derivative. In the social network example, the final unconditional probability is a product of the two probabilities at the two stages. An effect on, or change in, the probabilities can be expressed in the resulting cross-product as

$$(P_1 + \partial P_1)(P_{2|1} + \partial P_{2|1})$$

$$= P_1 \cdot P_{2|1} + (\partial P_1 \cdot P_{2|1} + P_1 \cdot \partial P_{2|1} + \partial P_1 \cdot \partial P_{2|1})$$

$$= P_f + \partial P_f, \qquad [4.6]$$

where P_f indicates the final probability. The partial derivative of P_f with respect to x_k in a sequential model with two stages is ∂P_f and has the components of the last parentheses. Typically in empirical research a sequential model seldom has more than three stages, and often a three-stage sequential model involves products of no more than two terms at a time (e.g., the car acquisition example above). Therefore, high school algebra will suffice for calculating partial derivatives in sequential models in most research settings.

Using the estimated probabilities that form the basis for Figures 4.1 and 4.2, two marginal effects of religious homogamy/heterogamy on event probabilities are calculated for the social network example: They are estimated at 12 years of completed education, with no educational difference but religious difference and with 2 years of educational difference and religious difference between the spouses while other variables are kept at their sample means. Thus for a married male who has 12 years of education and no educational but religious difference (with "Same Religion" = 0) in the marriage, P_1 is 0.463 and $P_{2|1}$ is 0.748, using the estimates and the means from Table 4.1 and Equation 4.4. The marginal effect of religious homogamy in the first stage for such a male is approximately 0.203, because $\partial P_1 = 0.463(1 - 0.463)0.818 = 0.203$; the marginal effect of religious homogamy in the second stage is approximately 0.259, because $\partial P_{2|1} = 0.748(1 - 0.748)1.374 = 0.259$. The marginal effect of religious homogamy on the final probability of naming spouse first is then approximately 0.324, because $\partial P_f = 0.203 \cdot 0.748 + 0.463 \cdot 0.259 + 0.203 \cdot 0.259 = 0.324$ (Equation 4.6), indicating that the event probability would increase roughly by 0.324 were he in religious homogamy.

Similarly, for a married male who has 12 years of education and 2 years of educational and religious difference (with "Same Religion" = 0) in the marriage, P_1 is 0.439 and $P_{2|1}$ is 0.726, using the estimates and the means from Table 4.1 and Equation 4.4. The marginal effect of religious homogamy in the first stage for such a male is approximately 0.201, because $\partial P_1 = 0.439(1 - 0.439)0.818 = 0.201$; the marginal effect of religious homogamy in the second stage is approximately 0.273, because $\partial P_{2|1} = 0.726(1 - 0.726)1.374 = 0.273$. The marginal effect of religious homogamy on the final probability is then approximately 0.321, because $\partial P_f = 0.201 \cdot 0.726 + 0.439 \cdot 0.273 + 0.201 \cdot 0.273 = 0.321$, indicating that the event probability would increase by 0.321 were he in religious homogamy. Again, partial derivatives for dummy variables, used here for illustrative purposes, are only a rough approximation (see Chapter 5). Taking the difference between corresponding predicted probabilities will provide the marginal effects as 0.263 and 0.265, respectively, showing that taking partial derivatives once again overestimates the marginal effect. The estimated marginal effect of religious homogamy on event probability is quite sizable, though it varies with educational difference in the marriage (and level of education, not examined here).

Next we examine the effect of education, a continuous variable on the final (unconditional) probability of naming spouse first as someone to include in the network. Because education interacts with educational difference between the partners, the result of Equation 4.5, similar to interpreting marginal effects in linear regression models with interaction terms, should be modified to $P(1 - P)(\hat{\beta}_k + \hat{\beta}_{in}x_o)$, in which $\hat{\beta}_{in}$ represents the parameter estimate for the interaction term, and x_o represents the other variable involved in the interaction. In the current example, the other variable is educational difference. Because our major concern here is homogamy/heterogamy, we choose to examine the marginal effect of education on probability for those from religious homogamy and those from religious heterogamy separately, with other variables held at their means. For simplicity in illustration the effect is calculated at 0 and 2 years of educational difference only. Using the modified version of Equation 4.5, the marginal effects of education at 0 and 2 years of educational difference in Stage 1 among religiously heterogamous males are approximately 0.011 and 0.017, because $\partial P_1 = 0.463(1 - 0.463)(0.046 + 0.012 \cdot 0) = 0.011$ and $\partial P_1 = 0.439(1 - 0.439)(0.046 + 0.012 \cdot 2) = 0.017$. The corresponding effects in Stage 2 are approximately -0.006 and 0.015, because $\partial P_{2|1} = 0.748(1 - 0.748)(-0.032 + 0.053 \cdot 0) = -0.006$ and $\partial P_{2|1} = 0.726(1 - 0.726)(-0.032 + 0.053 \cdot 2) = 0.015$. The marginal effects of education at 0 and 2 years of educational difference on the final probability then are approximately 0.005 and 0.019, because $\partial P_f = 0.011 \cdot 0.748 - 0.463 \cdot 0.006 - 0.011 \cdot 0.006 = 0.005$ and $\partial P_f = 0.017 \cdot 0.726 + 0.439 \cdot 0.015 + 0.017 \cdot 0.015 = 0.019$. Using the same method, I have also calculated the marginal effects of education for religiously homogamous males; they are estimated to be approximately 0.008 and 0.019 at 0 and 2 years, respectively, of educational difference. Thus, an additional year of education among religiously heterogamous males tends to increase the probability of naming spouse first as someone important in the social network by 0.005 if there is no educational difference between the spouses, and by 0.019 if the educational difference is 2 years. The same effect tends to be a bit stronger among religiously homogamous males if there is no educational difference, and stays about the same for those religious homogamous males who have a 2 year difference in education with their spouses.

This example of interpreting results from the social network example serves as a guide to interpreting marginal effect on event probability in sequential models comprised of a series of binary-outcome models. In the event that one (or more) of the stages involves another type of model, such as a multinomial logit model as in the example of premarital pregnancy and its resolution, extensions can be made to compute relevant marginal effects. The reader, however, should refer to other chapters, such as Chapter 6 on multinomial models, so that ideas from there can be combined to calculate marginal effect accordingly.

5. ORDINAL LOGIT AND PROBIT MODELS

The model I introduced in Chapter 4 deals with only one type of polytomous response—sequential response. Sometimes response categories are ordered but do not form an interval scale. Such responses are common in the social sciences. Attitudinal questions on social and public opinion surveys often take the form of Likert-type scales covering a range from "strongly disagree" to "strongly agree" or from "least important" to "most important." Other survey question patterns such as "never, sometimes, often, always" and "poor, adequate, good, very good, excellent" are also examples of responses with ordered categories. Other examples of ordinal responses in the social sciences include level of job skill, level of educational attainment, employment status (unemployed, part time, and full time). Such responses are normally coded 0, 1, 2, 3, and so on (or 1, 2, 3, 4, etc.). There is a clear ranking among the categories, but the differences among adjacent categories cannot be treated as the same.

Responses like these with ordered categories cannot be easily modeled with classical regression. Ordinary linear regression is inappropriate because of the noninterval nature of the dependent variable—the spacing of the outcome choices cannot be assumed to be uniform. On the other hand, multinomial logit models, the topic of the next chapter, though they could be used, would fail to account for the ordinal nature of the dependent variable and thus not employ all of the information available in that variable. Ordinal logit and probit models have been widely used for analyzing such data (see Maddala, 1983; McKelvey & Zavoina, 1975).

The Model

The model is another natural extension of the binary-outcome model, built around a latent regression in the same manner as the binary logit or probit model (the discussion here follows Greene, 1990):

$$y^* = \sum_{k=1}^{K} \beta_k x_k + \varepsilon . \qquad [5.1]$$

As with the binary-outcome model, y^* is unobserved and thus can be thought of as the underlying tendency of an observed phenomenon, and we assume that ε follows a certain symmetric distribution with zero mean such as the normal or logistic distribution. What we do observe is

$$
\begin{aligned}
y &= 1 & &\text{if } y^* \leq \mu_1 \,(=0), \\
&= 2 & &\text{if } \mu_1 < y^* \leq \mu_2, \\
&= 3 & &\text{if } \mu_2 < y^* \leq \mu_3, \\
&\ \ \vdots \\
&= J & &\text{if } \mu_{J-1} < y^*,
\end{aligned}
\qquad [5.2]
$$

where y is observed in J number of ordered categories, and the μs are unknown threshold parameters separating the adjacent categories to be estimated with βs. Consider, for example, if respondents were asked, "Is it important for the United States to keep a military presence in a war-torn country or area of the world such as Eastern Africa?" The respondents might vary in intensity of feeling about the question depending on certain measurable factors, xs, and certain unobserved factors, ε. In principle, they could respond to the questionnaire with their own y^* if we phrase the question as such. Usually given 3-7 ordered choices ("least important" to "most important" and some intermediate categories), they are forced to choose the category that most closely represents their own feelings on the question. That is why we sometimes find it hard to make a choice when answering a question with ordered responses, and may wish that there were other choices in between.

In general, we have

$$
\text{Prob}(y = j) = F\left(\mu_j - \sum_{k=1}^{K} \beta_k x_k\right) - F\left(\mu_{j-1} - \sum_{k=1}^{K} \beta_k x_k\right). \qquad [5.3]
$$

Equation 5.3 gives the general form for the probability that the observed y falls into category j, and the μs and the βs are to be estimated with an ordinal logit or probit model. Here we use a general cumulative distribution function, F, rather than specifying a particular form of distribution (which might be the logistic or the normal). In order for all the probabilities to be positive, we must have

$$0 < \mu_2 < \mu_3 < \ldots < \mu_{J-1}.$$

The first threshold parameter, μ_1, is typically normalized to zero so that we have one less parameter to estimate. This is feasible because the scale is arbitrary and can start or finish with any value. Without this normalization there would be $J - 1$ number of μs to estimate because the number of thresholds is always one smaller than the number of categories; with the normalization ($\mu_1 = 0$), there will be $J - 2$ number of μs to estimate. As before, either the probit or the logit link function can be easily used.

1. The Ordinal Probit Model. In the probit case, we have

$$\text{Prob}(y = 1) = \Phi\left(-\sum_{k=1}^{K} \beta_k x_k\right),$$

$$\text{Prob}(y = 2) = \Phi\left(\mu_2 - \sum_{k=1}^{K} \beta_k x_k\right) - \Phi\left(-\sum_{k=1}^{K} \beta_k x_k\right),$$

$$\text{Prob}(y = 3) = \Phi\left(\mu_3 - \sum_{k=1}^{K} \beta_k x_k\right) - \Phi\left(\mu_2 - \sum_{k=1}^{K} \beta_k x_k\right), \qquad [5.4]$$

$$\vdots$$

$$\text{Prob}(y = J) = 1 - \Phi\left(\mu_{J-1} - \sum_{k=1}^{K} \beta_k x_k\right).$$

It is easy to see that the second term in every line in Equation 5.4 is the corresponding cumulative standard normal distribution probability from the line above. Thus $\text{Prob}(y = 2) = \text{Prob}(y \le 2) - \text{Prob}(y \le 1)$. In general, we get $\text{Prob}(y = j)$ by taking the difference between two adjacent cumulative probabilities, obtained using ordinary probit models, with the exception of the first and the last category because $\text{Prob}(y \le 1) = \text{Prob}(y = 1)$ and $\text{Prob}(y \le J) = 1$. It is this logic that underlies both Equations 5.3 and 5.4.

2. *The Ordinal Logit Model.* In the logit case, we have

$$\log\left[\frac{P(y \leq j \mid x)}{1 - P(y \leq j \mid x)}\right] = \mu_j - \sum_{k=1}^{K} \beta_k x_k, \quad j = 1, 2, \ldots, J-1. \quad [5.5]$$

The only difference between the ordinal and the binary logit model is that the ordinal-outcome model allows a sequence of log-odds or logits specified with the same βs and xs but different μs. The left-hand side is also called a cumulative logit (Agresti, 1990), an alternative to continuation ratios (Fienberg, 1980). We can express the same logit relation in probability (some prefer to call this formulation logistic regression):

$$\text{Prob}(y \leq j) = \text{Prob}(y^* \leq \mu_j) = \frac{e^{\mu_j - \sum_{k=1}^{K} \beta_k x_k}}{1 + e^{\mu_j - \sum_{k=1}^{K} \beta_k x_k}}. \quad [5.6]$$

This gives us the counterpart of $\Phi(\cdot)$ according to the cumulative logistic distribution. We may call it $L(\cdot)$. Replacing $\Phi(\cdot)$ with $L(\cdot)$, we can express the ordinal logit model in terms of probability in a formulation similar to Equation 5.4:

$$\text{Prob}(y = 1) = L\left(-\sum_{k=1}^{K} \beta_k x_k\right),$$

$$\text{Prob}(y = 2) = L\left(\mu_2 - \sum_{k=1}^{K} \beta_k x_k\right) - L\left(-\sum_{k=1}^{K} \beta_k x_k\right),$$

$$\text{Prob}(y = 3) = L\left(\mu_3 - \sum_{k=1}^{K} \beta_k x_k\right) - L\left(\mu_2 - \sum_{k=1}^{K} \beta_k x_k\right), \quad [5.7]$$

.
.
.

$$\mathrm{Prob}(y = J) = 1 - L\left(\mu_{J-1} - \sum_{k=1}^{K} \beta_k x_k\right).$$

It is obvious that the difference between ordinal logit and probit models is only in their distribution functions. Both the ordinal logit and probit models can be estimated with SAS (the procedure PROC LOGISTIC) and LIM-DEP (the ordered probit procedure). The SAS procedure estimates ordinal models using one of the three link functions including the logit or the probit, but care needs to be taken when calculating probabilities because SAS writes, instead of Equation 5.5, $\log[\mathrm{Prob}(y \le j)/1 - \mathrm{Prob}(y \le j)] = \alpha_j + \beta_2 x_2 + \beta_3 x_3 + \ldots + \beta_k x_k$, where α_j is a composite term of μ_j and β_1, and "−" is replaced with "+" . This formulation, of course, will not affect maximum likelihood estimation, though users may be surprised to see that every coefficient bears a sign contrary to its substantive expectation.

An issue in the ordinal-outcome model is whether the β estimates are invariant to the thresholds. That is, the effects of an x should be constant regardless of the choice of response category j. This is called the parallel lines assumption. It is often called the proportional odds assumption in the logit case, and the equal slopes assumption in the probit case. The SAS procedure PROC LOGISTIC performs a test for the null hypothesis that the β effects are independent of response category j. For a more detailed discussion, see DeMaris (1992) and McCullagh and Nelder (1989).

Interpretation of Ordinal Logit and Probit Models

Because of the similarities between the ordinal logit and probit model, interpretation of the results from these models will be discussed together. I use results from an ordinal probit model to illustrate the three ways of interpretation, and supply the interpretation of ordinal logit estimates wherever applicable.

1. Marginal Effect on η or Transformed η. Because logit models lend themselves more easily to this way of interpretation, let us focus on interpreting logit estimates. If the above-mentioned proportional odds assumption is held, the interpretation of logit estimates is rather straight-forward, similar to those from a binary logit model. When the assumption holds, the partial effect of an x is invariant to the choice of response category j. Equation 5.5 implies that the estimated β should be the same

TABLE 5.1
Ordinal Probit Model Estimated for New Navy Recruits, $N = 5,641$

x Variable	$\hat{\beta}$	t Ratio	$\bar{x}$
Tech Training Guarantee?	0.057	1.7	0.66
Mom's Education	0.007	0.8	12.1
AFQT Score	0.039	39.9	71.2
Education Attainment	0.190	8.7	12.1
Married When Enlisted?	−0.480	−9.0	0.08
Age When Enlisted	0.0015	0.1	18.8
Constant	−4.340	—	—
$\hat{\mu}$	1.79	80.8	—

SOURCE: Greene (1990, Table 20.12). Reprinted with the permission of Macmillan College Publishing Company from ECONOMETRIC ANALYSIS, Second Edition, by William H. Greene. Copyright © 1993 by Macmillan College Publishing Company.
NOTE: Dummy variables are indicated with question marks, and coded 1 if "yes," 0 otherwise.

regardless of which j is of concern. The marginal effect of x_k in a binary case is interpreted as the expected change in the odds of belonging to category 1 rather than 2, which is the multiplicative effect of $\exp(\beta)$, given a unit change in x_k. Because there are only two categories, Prob($y \leq 1$) is equal to Prob($y = 1$). This, however, is no longer true for a response with three or more categories. For a response with three categories, Equation 5.5 suggests that the effect of x_k would induce a change in the odds of responding to category 1 instead of 2 and 3, or 1 or 2 instead of 3, by a factor of $\exp(\beta)$. The contrast is always between the probability of belonging to the first up to the jth category and the probability of belonging to the remaining categories. An example will illustrate the usage.

Greene (1990) discusses an ordinal probit model estimated for the job assignments of new Navy recruits. The dependent variable is an ordered response with three categories—job assignments of new Navy recruits classified as "medium skilled," "highly skilled," and "nuclear qualified/highly skilled." The determinants include (a) a dummy variable of whether the recruit entered the Navy with an "A school" (technical training) guarantee, (b) educational level of the individual's mother, (c) score on the Air Force Qualifying Test (AFQT), (d) completed years of education by the entrant, (e) a dummy variable of whether the trainee was married when enlisted, and (f) individual's age when recruited. An ordinal probit model was estimated and its results are reproduced in Table 5.1. Two estimates have extremely large t ratios: the β for the AFQT score and μ. The AFQT score is a primary sorting device for assigning job classifications in the

Navy. A highly significant, positive μ estimate indicates that the three categories in the response are indeed ordered. There is only one μ estimate because $J - 2 = 3 - 2 = 1$ with the first μ normalized to be zero.

Let us examine the effect of two determinants—a dummy variable and a continuous variable. First, let us look at the effect of marital status at the time of enlistment on job classifications. The logit estimate for marital status is approximately -0.768.[2] Exponentiating it gives 0.464, which is the estimated effect on the odds. This marginal effect suggests that the odds for those who were married when enlisted to have been classified as highly or nuclear qualified/highly skilled instead of medium skilled are about 0.464 times as high as those who were not married, other things being equal. Similarly, the odds for the married recruits to have been classified as nuclear qualified/highly skilled instead of medium or highly skilled are about 0.464 times as high as those not married. The negative sign in Equation 5.5 ensures that large values in β_k lead to an increase of probability in the higher numbered categories (McCullagh & Nelder, 1989) and an increased odds of falling into higher numbered categories despite that the cumulative logit contrasts the lower with the higher numbered categories. Note that one needs to multiply the coefficient from SAS by -1 before interpretation unless one intentionally reverse-codes the dependent variable prior to estimation, due to the difference in formulation discussed earlier.

The effect of education can be interpreted in a similar way. The logit estimate for the trainee's own education is about 0.304, and the corresponding effect on the odds after exponentiation is 1.355. Other things being equal, the odds of being classified as highly skilled or nuclear qualified/highly skilled versus less skilled would be 1.355 times greater with a one-year increase in the trainee's education. Under the same circumstances, the odds of being classified as nuclear qualified/highly skilled versus less skilled would be 1.355 times greater with a one-year increase in the trainee's education. Therefore, interpreting marginal effects on transformed η remains an easy, flexible, and useful option for making sense of logit models with ordered responses.

2. Predicted Probabilities Given a Set of Values in the Explanatory Variables. The example of Navy recruits has only three categories in the response variable; Equation 5.4 thus simplifies to

$$\text{Prob}(y = 1) = \Phi\left(-\sum_{k=1}^{K} \beta_k x_k\right),$$

$$\text{Prob}(y=2) = \Phi\left(\mu - \sum_{k=1}^{K} \beta_k x_k\right) - \Phi\left(-\sum_{k=1}^{K} \beta_k x_k\right), \qquad [5.8]$$

$$\text{Prob}(y=3) = 1 - \Phi\left(\mu - \sum_{k=1}^{K} \beta_k x_k\right).$$

With three response categories, we need to estimate one μ parameter only. ($J - 2 = 3 - 2 = 1$.) As a result, we can drop the subscript of μ without causing confusion. We can imagine the three response probabilities to be the area under a normal curve partitioned into three regions by the two thresholds. The first threshold is always normalized to zero minus the influence of xs; the second is the estimated μ minus the influence of xs. Because $\Phi(\cdot)$ gives the standard normal cumulative probability, the area in region 1 is given by $\Phi(0 - \sum_k \hat{\beta}_k x_k) = \Phi(-\sum_k \hat{\beta}_k x_k)$, as implied by Equation 5.3 and expressed in the first line of Equation 5.8. The area in region 2 equals the area up to the point of the estimated μ minus the influence of xs with the area in region 1 subtracted, hence line 2 in Equation 5.8. Finally, the area in region 3 equals the probability of the total area, which is 1, minus the area up to the point of the estimated μ minus the influence of xs (or the sum of regions 1 and 2), hence line 3 in Equation 5.8.

To facilitate interpretation, we now calculate the predicted probabilities at five levels of completed education, with other x variables valued at their means. The results are reported in Table 5.2. One may, if one wishes, graph these data, as we did for the example in Chapter 4. These predicted probabilities give a more detailed picture of the effect of education than simply interpreting the marginal effect on the transformed η. The results indicate that an increased level of education will reduce a recruit's probability of being classified as medium skilled; increased education will increase the individual's probability of being classified as highly skilled if the person has less than a high school education, while reducing that probability if the person has at least a high school diploma; and that increased education will increase the probability of being classified as nuclear qualified/highly skilled, with each two additional years more or less doubling the probability. One can, of course, use specific levels of education and values of predicted probability to discuss the interpretation.

Another way to describe the change in predicted probabilities as education increases is to look at the probability distribution at particular levels

TABLE 5.2
Predicted Probabilities of Job Classifications

Level of Education	Level of Skill		
	Medium Prob(y = 1)	Highly Prob(y = 2)	Nuclear Prob(y = 3)
8 years	0.473	0.485	0.043
10 years	0.327	0.583	0.090
12 years	0.204	0.628	0.168
14 years	0.113	0.606	0.281
16 years	0.056	0.524	0.420

of education. It is obvious that at a low level of education, a recruit is as likely to have been classified as medium skilled as to have been classified as highly skilled. As education increases, a certain amount of probability gets shifted out of Prob($y = 1$) into Prob($y = 2$) and consequently out of Prob($y = 2$) into Prob($y = 3$). For a college graduate, the pattern is almost reversed: A college graduate trainee is almost as likely to have been classified as nuclear qualified/highly skilled as to have been classified as highly skilled. On the other hand, the number of college graduates or high school dropouts in the sample possibly will not be extremely high. Therefore, since the sample mean education is 12.1, the predicted probabilities at 12 years education should give an estimate close to the observed percentages in job classifications.

One point to note here is that the three predicted probabilities across the response categories at a particular level of education should sum up to 1. The equality serves as a check for verifying the correctness of the calculation. A quick browse will reveal that all rows except one add up to exactly 1; the exception, row 1, is due to rounding error. It is especially important to verify the results of calculation if there are many variables, many points of prediction, or both. A spreadsheet software will facilitate the calculation of predicted probabilities tremendously.

3. Marginal Effect on the Probability of an Event. As with binary or sequential logit and probit models, we express the marginal effect on event probability in ordinal logit and probit models as the partial derivative of probability with respect to x_k. In general, we have

$$\frac{\partial \text{Prob}(y = j)}{\partial x_k} = \left[f\left(\mu_{j-1} - \sum_{k=1}^{K} \beta_k x_k \right) - f\left(\mu_j - \sum_{k=1}^{K} \beta_k x_k \right) \right] \beta_k, \quad [5.9]$$

TABLE 5.3
Marginal Effect on the Probabilities of Job Classifications

Statistic	$-\sum_k \hat{\beta}_k x_k$	$\hat{\mu} - \sum_k \hat{\beta}_k x_k$	Prob(y = 1)	Prob(y = 2)	Prob(y = 3)
AFQT Score = $\bar{x}$	−0.8479	0.9421	0.1982	0.6287	0.1731
AFQT Score = $\bar{x} + 1$	−0.8869	0.9031	0.1876	0.6292	0.1832
Change			−0.0106	0.0005	0.0101
$\partial P / \partial x_k$			−0.0109	0.0009	0.0100
Marital Status = 0	−0.8863	0.9037	0.1877	0.6292	0.1831
Marital Status = 1	−0.4063	1.3837	0.3423	0.5745	0.0832
Change			0.1546	−0.0547	−0.0999
$\partial P / \partial x_k$			0.1293	−0.0020	−0.1273

where $f(\cdot)$ represents probability density function, such as the standard normal or logistic. For the example, we can specify $f(\cdot)$ to be $\phi(\cdot)$, and simplify Equation 5.9 down to one with three response categories:

$$\frac{\partial \text{Prob}(y = 1)}{\partial x_k} = -\phi\left(\sum_{k=1}^{K} \beta_k x_k\right)\beta_k ,$$

$$\frac{\partial \text{Prob}(y = 2)}{\partial x_k} = \left[\phi\left(-\sum_{k=1}^{K} \beta_k x_k\right) - \phi\left(\mu - \sum_{k=1}^{K} \beta_k x_k\right)\right]\beta_k ,$$

$$\frac{\partial \text{Prob}(y = 3)}{\partial x_k} = \left[\phi\left(\mu - \sum_{k=1}^{K} \beta_k x_k\right)\right]\beta_k . \qquad [5.10]$$

Again, variables other than the one being interpreted are held at their mean values. In Chapters 3 and 4, I mentioned that partial derivatives for dummy variables are in principle inaccurate and thus may differ in value from assessing the change in predicted probability. In the following example, we study the marginal effect on probability of two variables—AFQT score and marital status, one continuous and one dummy (Table 5.3). The effect of each variable is assessed in two ways—by taking the partial derivative and by computing the change in predicted probability given a unit change in x_k.

In the first two columns, values of $-\sum_k \hat{\beta}_k x_k$ and $\hat{\mu} - \sum_k \hat{\beta}_k x_k$ are reported to facilitate the reader with the intermediate steps in the calculation.

Probabilities are calculated using Equation 5.8, and the marginal effects on probability are calculated using Equation 5.10. The effect of AFQT score is quite distinct. With one point of increase in the score, the probability of being classified as medium skilled will decrease about 0.01, and that of being classified as nuclear qualified/highly skilled will increase about the same amount. The probability of being classified as highly skilled remains almost unchanged, increasing by a negligible amount. This marginal effect can be derived regardless of the method used. In fact, the difference between the "change" estimates and the partial derivatives is no greater than 0.0004 in any comparison. It shows that with a finely measured continuous variable such as AFQT score, either partial derivatives or differences in predicted probabilities should give approximately identical results.

With dummy variables, it is a different matter. As interpretation of the effect on the odds revealed, being married at the time of enlistment will increase the chance of being classified as medium skilled and decrease the chance of being classified as highly skilled, especially nuclear qualified/highly skilled. Using the change-in-probability method, the increase in the probability of being in category 1 is about 15%, while the reduction in the probability of being in category 3 is about 10%. Using partial derivatives, the increase is about 13%, and the decrease is also about 13%. The reduced probability of being classified as highly skilled differs by about 5% between the two methods. The bias when using partial derivatives for dummy variables is noticeable. One may argue, however, that substantive findings will not change much because of the deviation in results. Since calculating predicted probabilities and then taking the difference to derive the change in probability is relatively straightforward for dummy variables, I recommend use of this method for the researcher interested in a more accurate description of the marginal effect of a dummy variable on event probability, and partial derivatives be used only for an overall impression and with caution.

That predicted probabilities sum up to unity implies a zero-sum game for effects on probability. Marginal effects on probability should sum up to zero by canceling one another out across the response categories. A quick eyeballing both of the marginal effect of AFQT score and that of marital status on event probability calculated by either method demonstrates that this condition holds true for the example in Table 5.3. This condition serves as a check for verifying the correctness of the results. The conditions that probabilities should sum to unity and that marginal effects should sum out to zero are both necessary but not sufficient conditions for verifying the accuracy of calculated results.

6. MULTINOMIAL LOGIT MODELS

In the polytomous-response models discussed in Chapters 4 and 5, there exists either an intrinsic order or a natural sequence among the response categories. In other polytomous-response models, the categories in the dependent variable are truly discrete, nominal, or unordered. When the data are, or are believed to be, of this type, a multinomial logit model is appropriate. The probit counterpart of a multinomial logit model involves solving multiple integration related to the multivariate normal distribution, and thus is computationally difficult in estimation and rarely used. Therefore, this chapter focuses on interpreting estimates from multinomial logit models.

Examples of unordered responses are many: They include choice among consumer products, occupations, academic majors and programs, religions, modes of transportation, and political candidates. Sometimes we are not sure if the categories are ordered or sequential in the response. If unsure, a multinomial logit model should be used. The rationale is that we should go with a statistical model that requires fewer or weaker assumptions if we are not sure the data satisfy all the assumptions specified by an alternative model. The researcher may also use a statistic such as the Akaike Information Criterion (AIC) to choose between two competing models that are not nested (see, e.g., Amemiya, 1981).

The Model

This model is another natural extension of the binary logit model. The multinomial logit model estimates the effects of explanatory variables on a dependent variable with unordered response categories: The equation,

$$\text{Prob}(y = j) = \frac{e^{\sum_{k=1}^{K} \beta_{jk} x_k}}{1 + \sum_{j=1}^{J-1} e^{\sum_{k=1}^{K} \beta_{jk} x_k}}, \qquad [6.1]$$

gives $\text{Prob}(y = j)$ where $j = 1, 2, \ldots, J - 1$. Note that parameters β have two subscripts in the model, k for distinguishing x variables, and j for distinguishing response categories. The subscript j indicates that now there are $J - 1$ sets of β estimates. In other words, the total number of parameter estimates will be $(J-1)K$. This implies that the sample size should be larger than $(J-1)K$. For 10 x (explanatory) variables including the intercept and 5 response categories in the dependent variable, the total number of parameter estimates will be 40. Most often the last response category is used as the reference category against which other response categories are compared. The reader will notice that if $J = 2$, Equation 6.1 simplifies to Equation 3.5 for binary logit models. The equation,

$$\text{Prob}(y=j) = \cfrac{1}{1 + \sum_{j=1}^{J-1} e^{\sum_{k=1}^{K} \beta_{jk} x_k}}, \qquad [6.2]$$

gives $P(y = J)$.[3] Alternatively, the last probability can also be derived by taking $1 - [\text{Prob}(y = 1) + \ldots + \text{Prob}(y = J - 1)]$.

The similarities between multinomial logit models and binary logit models can be seen no more clearly than in the multinomial model expressed in logit form. Equations 6.1 and 6.2 imply the following:

$$\log\left[\frac{\text{Prob}(y=j)}{\text{Prob}(y=J)}\right] = \sum_{k=1}^{K} \beta_{jk} x_k . \qquad [6.3]$$

For the connection between Equation 6.3 and Equation 6.1 or 6.2, see Maddala (1983). Obviously, when $J = 2$, Equation 6.3 simplifies to Equation 3.4, the binary logit model. Equation 6.3 should also remind us of the multinomial logit link function,

$$\eta_j = \log(\mu_j/\mu_J) .$$

The striking similarities in formulation between the binary logit and the multinomial logit model suggest several things. First, probability in a multinomial logit model can be calculated similarly to that in a binary logit model, with the only modification being accounting for multiple sets of β estimates. In addition, the meaning of logit (log-odds) and odds terms is identical in both models. In the binary case, the comparison is between category 1 and category 2 (or the first vs. the last category). In the multinomial case, the comparison is between category j and J (or any category but the last versus the last). These similarities will become transparent in the interpretation section. The multinomial logit model can be estimated using the PROC CATMOD (or PROC MLOGIT) procedure in SAS and the logit model procedure in LIMDEP.

One important issue in the use of multinomial logit models is the assumption of independence from irrelevant alternatives, or IIA. It is probably the most widely discussed aspect in the methodological literature on multinomial models. Simply stated, the IIA property holds that the ratio of the choice probabilities of any two alternatives (in response categories) for a particular observation is not influenced systematically by any other alternatives. A frequently cited example is the red bus/blue bus paradox. The commuter has two choices of modes of transportation—car and bus. The choice probability for each mode is 1/2 and the ratio of the two choice probabilities is 1. Suppose that another bus service is introduced that is equal in all attributes but different in color. Buses in one service are painted red and those in the other, blue. If the ratio of choice probabilities are to be constant, every choice probability should equal 1/3. But this is unrealistic because commuters are most likely to treat the two bus services as the same. That is, the choice probability for car is still 1/2, and the probability for either red or blue bus is 1/4. Now the ratio of the choice probability for car and that for a bus service is 2 instead of 1; thus the assumption of IIA is violated because some of the choices are not independent from each other. A more detailed discussion of this important topic is beyond the scope of the current monograph. Interested readers should refer to Ben-Akiva and Lerman (1985), Greene (1990), Train (1986), and Wrigley (1985). We should emphasize, however, that this is an important assumption, that its suitability needs to be considered carefully in each application of the multinomial logit model defined in this section, and that the above authors discuss both the consequence of violation of IIA for the estimates and procedures for testing the empirical plausibility of the assumption.

Interpretation of Multinomial Logit Models

Because of the similarities between the binary and the multinomial logit models, the three ways of interpretation resemble those for interpreting results from binary logit models. As noted earlier, a major difference is that the reference category now is no longer the other choice as in binary logit, but another category, typically the last. (The choice of the reference category is irrelevant for estimation and should be determined on substantive grounds.) Thus, when we interpret, we must adjust accordingly.

Whereas you compare the likelihood of event A with that of event B or non-A in a binary logit model, now you compare the likelihood of event A with that of event D, that of event B with that of event D, and that of event C with that of event D in a multinomial logit model with four categories in the dependent variable. As implied by Equation 6.1, the model gives three nonredundant sets of parameter estimates, each associated with one of the first three alternatives, and this is the usual computer output from statistical packages. For other contrasts such as A with B, A with C, and B with C, you may simply rearrange the sequence of the response categories and obtain the corresponding estimates by reestimating the model, unless you are skilled in computer programming and write a subroutine for computing these estimates.

1. Marginal Effect on η or Transformed η. Marginal effect on the odds for the multinomial logit model refers to the partial effect on the odds of falling into a category as opposed to a user-chosen reference category. As an example, Table 6.1 presents parameter estimates representing all contrasts among four categories of sterilization among U.S. white women surveyed in 1982. For detailed information about sample and substantive issues, see Rindfuss and Liao (1988). The model was estimated with the four types of sterilization arranged in the order of "sterilized for a contraceptive reason," "not sterilized," "sterilized for a medical reason," and "sterilized for mixed reasons." Of course the order does not assume any meaning. Instead of giving the three sets of parameters only contrasting the first three choices with the last, I present all six sets of estimates contrasting between the response categories (Table 6.1).[4] It is useful to present all contrasts sometimes because of substantive interest.

First, let us interpret the effect of a dummy variable. The variable of marital status has an estimate of −2.80 for the contrast of contraceptive versus no sterilization. Exponentiating it, we get 0.061. The odds for never-married women to have had a sterilizing operation for contraceptive purposes instead of having not been sterilized are only about 0.061 times as high as the same odds for married women. Now let us move to the comparison of contraceptive and medical sterilization. The parameter estimate for women not graduated from high school is −1.28, suggesting that the odds for them to have had sterilization for contraceptive purposes instead of for medical purposes are only about 0.278 times as high as the same odds for women with a high school diploma only.

There are only two continuous variables in the model, age and parity (number of children ever born), and both involve quadratic terms. This

TABLE 6.1
Multinomial Logit Model Results
for Type of Sterilization for White Women

x Variable	Contraceptive vs. Mixed	No Sterilization vs. Mixed	Medical vs. Mixed	Contraceptive vs. Medical	No Sterilization vs. Medical	Contraceptive vs. No Sterilization
Age	−0.12	−0.61*	−0.18	0.07	−0.43	0.49*
Age Squared	0.00	0.01	0.00	−0.00	0.01	−0.01*
Age at Menarche	0.01	0.01	0.04	−0.03	−0.03	−0.00
Parity in 1977	1.05**	−1.04**	0.13	0.92*	−1.16**	2.08**
Parity in 1977 Squared	−0.29**	0.16**	0.02	−0.31**	0.15*	−0.46**
Education in Years						
< 12	−0.66*	−0.09	0.61	−1.28**	−0.71*	−0.57*
13-15	−0.16	−0.04	−0.09	−0.07	0.05	−0.12
16 +	0.03	0.36	−0.11	0.15	0.48	−0.33
Catholic?	−0.25	0.03	−0.54	0.29	0.58	−0.28
Never-married?	−0.41	2.39*	1.81	−2.22	0.58	−2.80**
Region						
Northeast	−0.50	−0.03	−0.25	−0.26	0.22	−0.47
North Central	−0.24	−0.33	−0.43	0.19	0.10	0.09
West	−0.11	0.17	−0.41	0.30	0.58	−0.28
Metropolitan Residence	0.02	0.09	0.13	−0.11	−0.04	−0.06
Model χ^2	560.88					
N	2,520					

SOURCE: Reproduced with the permission of the Population Council, from Ronald R. Rindfuss and Futing Liao, "Medical and contraceptive reasons for sterilization in the United States," *Studies in Family Planning* 19, no. 6 (November/December 1988): 370-380.
NOTE: High school graduate is the reference category in education, and South is the reference category in region. Other dummy variables are indicated with question marks, and coded 1 if "yes," 0 otherwise.
* and ** indicate statistical significance at the .05 and .01 level, respectively.

means the effect on the odds varies with the level of the variable involved in the quadratic term. Such a marginal effect on the odds can be assessed by $\exp(\beta_2 + 2\beta_3 x_2)$, where x_2 is the variable involved in the quadratic term, β_2 is its main effect, and β_3 is the effect for the squared x_2.[5] For women aged 20, 25, 30, and 35, we have four corresponding marginal effects of 1.094, 0.990, 0.896, and 0.811 for the contrast between contraceptive and no sterilization. For women aged 20, the odds of having had a sterilizing operation for contraceptive purposes instead of having not been sterilized will be increased by 1.094 times with a one-year increase in age. However, for women aged 25, 30, and 35, such odds will be decreased by a factor of

0.990, 0.896, and 0.811, respectively, with a one-year gain in age. Similarly, we can interpret the marginal effect of parity on the odds of having had sterilization for a contraceptive reason as opposed to for mixed reasons. For women who have had 0, 2, 4, or 6 children, the corresponding marginal effects on the odds are 2.858, 0.896, 0.281, or 0.088. This suggests that for a woman of parity 0, the odds of having had sterilization for contraceptive purposes instead of for mixed reasons will be increased 2.858 times once she has a child. For women of parity of 2, 4, and 6, however, the same odds will be decreased by a factor of 0.856, 0.281, and 0.088, respectively, with the gain of an additional child. As in linear models, quadratic functions in multinomial logit models also describe nonlinear effects.

The IIA property mentioned in the previous section implies that the odds ratio of one choice to another does not depend on the other choices. From the point of view of estimation, it is useful that the odds ratio does not depend on the other choices. This follows from the independence of disturbances in the specification of the multinomial logit model. From a behavioral viewpoint, this may not be so attractive at times. Interested readers are referred especially to discussion in Ben-Akiva and Lerman (1985) as well as Greene (1990), Train (1986), and Wrigley (1985).

2. *Predicted Probabilities Given a Set of Values in the Explanatory Variables.* Estimated probabilities in multiple-outcome models should be even more useful than those in binary-outcome models, for instead of just one schedule of nonredundant probabilities now there are at least two schedules of nonredundant probabilities. The sterilization example has three schedules of nonredundant probabilities for the four response categories. The researcher probably wants to present all the schedules of predicted probabilities because with more than two categories it becomes more difficult to calculate $[1 - (P_1 + P_2 + \ldots + P_J)]$ in one's head.

For calculating the predicted probability of a sterilizing operation by parity for ever-married white women, the first three columns of estimates in Table 6.1 are used with Equations 6.1 and 6.2. As with previous chapters, variables not being interpreted are kept at their sample means. The calculation works out the same way as with the binary logit model, except that now the denominator involves summing over a series of exponentiated results. To illustrate the point more easily without going through the real-world example of sterilization, let us use a hypothetical case having a dependent variable with three response categories ($J = 3$) and two independent variables, x_2 and x_3 (x_1 being equal to 1). The variable x_2 is continuous, and x_3 is dichotomous. We want to calculate the predicted probability for the three response categories at $x_2 = 20$, 30, and 40 with x_3

held at its mean value of 0.6. The β_{1k} and β_{2k} are estimated from the multinomial logit model with the reference category $J = 3$. The equation for the first response category is

$$\text{Prob}(y = 1 \mid x_2 = 20, x_3 = 0.6)$$

$$= \frac{e^{\beta_{11} \cdot 1 + \beta_{12} \cdot 20 + \beta_{13} \cdot 0.6}}{1 + e^{\beta_{11} \cdot 1 + \beta_{12} \cdot 20 + \beta_{13} \cdot 0.6} + e^{\beta_{21} \cdot 1 + \beta_{22} \cdot 20 + \beta_{23} \cdot 0.6}}. \quad [6.4]$$

This gives us the predicted probability of choice 1 with $x_2 = 20$ and $x_3 = 0.6$. For probabilities at $x_2 = 30$ and $x_2 = 40$, simply replace the value 20 with 30 and 40 and go over Equation 6.4 two more times. The corresponding predicted probability of choice 2 is calculated as

$$\text{Prob}(y = 2 \mid x_2 = 20, x_3 = 0.6)$$

$$= \frac{e^{\beta_{21} \cdot 1 + \beta_{22} \cdot 20 + \beta_{23} \cdot 0.6}}{1 + e^{\beta_{11} \cdot 1 + \beta_{12} \cdot 20 + \beta_{13} \cdot 0.6} + e^{\beta_{21} \cdot 1 + \beta_{22} \cdot 20 + \beta_{23} \cdot 0.6}}. \quad [6.5]$$

The reader will notice that the only difference between Equations 6.4 and 6.5 is in the subscripts of β in the numerator. In Equation 6.4 the first set of parameter estimates are used while in Equation 6.5 the second set of estimates are used. Finally, the predicted probability of the last response category is

$$\text{Prob}(y = 3 \mid x_2 = 20, x_3 = 0.6)$$

$$= \frac{1}{1 + e^{\beta_{11} \cdot 1 + \beta_{12} \cdot 20 + \beta_{13} \cdot 0.6} + e^{\beta_{21} \cdot 1 + \beta_{22} \cdot 20 + \beta_{23} \cdot 0.6}}. \quad [6.6]$$

Models with variables calculated at more values in x variables and with more response categories in the dependent variable are simply extensions of this example. A spreadsheet software will come in handy for calculating these probabilities.

Returning to the sterilization example, the probabilities of having not had sterilization and of having had sterilization for a contraceptive reason, a medical reason, and mixed reasons by parity for ever-married white

TABLE 6.2

Predicted Probability of a Sterilizing Operation by Parity,
for Ever-Married White Women

| Parity | Not Sterilized | Percentage | | |
| | | Sterilized | | |
		Contraceptive	Medical	Mixed
0	91	4	4	1
1	72	18	8	2
2	57	29	10	4
3	58	23	13	6
4	69	9	14	8
5	80	1	10	8
6	89	0	5	6

SOURCE: Reproduced with the permission of the Population Council, from Ronald R. Rindfuss and Futing Liao, "Medical and contraceptive reasons for sterilization in the United States," *Studies in Family Planning* 19, no. 6 (November/December 1988): 370-380.
NOTE: "Sterilizing operation" includes those performed on either the woman or her partner. Some rows may not add up to 100 due to rounding error.

women are presented in Table 6.2. The probabilities are presented in percentages instead of proportions as in earlier chapters. For many people percentages are easier to read.

Without going into the substantive conclusions from these predicted probabilities, a few quick general observations are in order. The estimated probability for no sterilization is extremely high for women with no child. It decreases for women with two children, and then increases with parity. The predicted probability of having had sterilization for a contraceptive reason is small for women with no child, peaks for women with two children, and drops down to a very low level for women with more than four children. The estimated probability is relatively small for women having had sterilization either for a medical reason or for mixed reasons. Other things being equal, women with three or four children tend to have a higher probability of having had sterilization for a medical reason, but women with four or five children tend to have a higher probability of having had sterilization for no clear medical or contraceptive reasons.

3. Marginal Effect on the Probability of an Event. The interpretation of marginal effect of an independent variable on choice probability in a multinomial logit model resembles that in a binary logit model, but represents a more general case because of the flexible number of response categories. The meaning of the marginal effect, however, still refers to a particular response category.

Again, we take partial derivative of a response probability with respect to x_k. When there are J number of response categories, the partial derivative of Prob($y = j$) with respective to x_k is

$$\frac{\partial \text{Prob}(y=j)}{\partial x_k} = P_j \left(\beta_{jk} - \sum_{j=1}^{J-1} P_j \beta_{jk} \right), \qquad [6.7]$$

where P_j is short for Prob($y = j$). If there are only two response categories ($J = 2$), Equation 6.7 reduces to Equation 3.12 for marginal effect on probability in binary logit models.

As discussed in previous chapters, the marginal effect of a dummy variable on event probability calculated by taking derivatives is in principle inaccurate, though it can give a rough approximation. As with other probability models discussed so far, the marginal effect of a dummy variable on event probability can always be accurately derived by taking the difference between the predicted probability when the variable is equal to 1 and that when it is equal to 0. Thus we will focus on the interpretation of a continuous variable in the sterilization example. Two variables in the analysis can be treated as continuous—age and parity. Let us examine the effect of parity simply because it has significant estimates in more contrasts.

The quadratic function of parity necessitates a version of Equation 6.7 modified by having a quadratic function of x_k replace the simple linear function of x_k. From the interpretation of its effect on the odds, we should know how to modify it (also see Notes 3 and 5). The marginal effect of a variable, x_k, involved in a quadratic term on event probability is given by

$$\frac{\partial \text{Prob}(y=j)}{\partial x_k} = P_j \left[(\beta_{jk} + 2\beta_{jk+1} x_k) - \sum_{j=1}^{J-1} P_j (\beta_{jk} + 2\beta_{jk+1} x_k) \right], \quad [6.8]$$

where β_{jk} represents the main effect of x_k, and β_{jk+1} indicates the parameter estimate for squared x_k. Equation 6.8 gives marginal effect of the quadratic function of x_k on event probability; in general, it also implies how to solve for marginal effect of other nonlinear functions. First, derive the relevant partial derivative with respect to x_k as if it were in a linear model. Then plug the partial derivative in the two inner parentheses in Equation 6.8. The result will be the partial derivative of event probability with respect to x_k.

TABLE 6.3

Marginal Effect of Parity on Probability of a Sterilizing Operation

| | | Percentage | | |
| | | | Sterilized | |
Parity	Not Sterilized	Contraceptive	Medical	Mixed
0	−12.8	7.8	4.1	0.9
1	−21.6	16.0	4.7	0.8
2	−9.2	3.7	4.5	1.0
3	5.4	−11.9	5.5	1.0
4	10.2	−12.3	2.8	−0.7
5	7.8	−2.3	−1.3	−3.7
6	7.0	−0.0	−2.2	−4.8

NOTE: "Sterilizing operation" includes those performed on either the woman or her partner. Some rows may not add out to 0 due to rounding error.

With four response categories in the dependent variable in the sterilization example, we need to go through Equation 6.8 four times. The predicted probabilities in Table 6.2 are used in the calculation of marginal effects. The corresponding estimates are plugged in Equation 6.8 as β_{jk} and β_{jk+1} for $j = 1, 2$, and 3, respectively, to derive corresponding partial derivatives (Table 6.3).[6] For instance, to compute the first entry (−0.128 or −12.8%) in Table 6.3, the marginal effect of parity (at parity 0) on the probability of not being sterilized, we have

$$\frac{\partial \text{Prob}(y = 1)}{\partial \text{Parity}}$$

$$= P_1 \left[(\beta_{1,\,\text{parity}} + 2\beta_{1,\,\text{parity}^2} \cdot \text{Parity}) - \sum_{j=1}^{J-1} P_j(\beta_{j,\,\text{parity}} + 2\beta_{j,\,\text{parity}^2} \cdot \text{Parity}) \right],$$

$$= 0.91(-1.04 + 2 \cdot 0.16 \cdot 0)$$

$$- 0.91[0.91(-1.04 + 2 \cdot 0.16 \cdot 0) + 0.04(1.05 + 2 \cdot -0.29 \cdot 0)$$

$$+ 0.04(0.13 + 2 \cdot 0.02 \cdot 0)]$$

$$= -0.128. \quad [6.9]$$

The other marginal effects in Table 6.3 are derived accordingly.

The four schedules of marginal effects of parity in 1977 on the probability of having had a sterilizing operation by 1982, presented in Table 6.3, are quite telling. Overall, the marginal effect on the probability of remaining not sterilized will be negative when parity is low and positive when parity is high, but the effect is actually a complex changing function of parity. This complexity is because we have three $(J - 1 = 4 - 1 = 3)$ independent event probabilities in the model. For women with one child, the probability of having had contraceptive sterilization will increase by approximately 16 percentage points with an additional child; for a woman with three children, this probability will decrease by about 12 percentage points with the additional child. An additional child tends to increase a woman's probability of having had sterilization for medical purposes if she is of low parity, and decrease the probability if she is of parity five or higher. If Tables 6.2 and 6.3 are compared, we will notice that the schedules of marginal effects by and large correspond to the differences in adjacent parities in the schedules of predicted probabilities. The discrepancies in magnitude are in part due to the debatable continuous nature of the parity variable (with a limited number of intervals) and in part due to the small number of significant digits displayed in the original article and hence used in the calculation of marginal effect.

Alternatively, we may wish to graph marginal effects such as those in Table 6.3, just as we sometimes may present predicted probabilities graphically. The graph in Figure 6.1 gives a more direct understanding of the marginal effect of parity on the four response probabilities. It also visually shows that the marginal effect interpretation is a zero-sum game across response categories in multinomial logit models, as in binary- and ordinal-outcome models. To check if the zero-sum condition holds, we still need to go back to Table 6.3 and add up the values across columns. A quick examination will reveal the condition holds except for rounding errors.

In sum, interpreting a parameter estimate for an explanatory variable in terms of its effects on event probability in the multiple-outcome model demonstrates clearly that a variable has not just one but $J - 1$ independent effects on $J - 1$ event probabilities. The point is shown by all three of the interpretation methods in this chapter. We interpret the effect of a variable on the odds of taking on category j versus J; we present J schedules of estimated probabilities; we calculate J schedules of marginal effects of a variable on event probabilities. The same features will be found in another probability model dealing with multiple response categories to be discussed in Chapter 7—the conditional logit model.

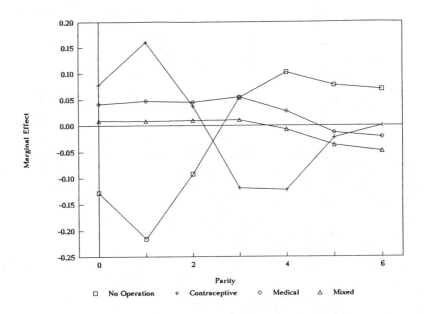

Figure 6.1. Marginal Effect of Parity on Probability of Sterilization

7. CONDITIONAL LOGIT MODELS

A variation of the multinomial logit model discussed in Chapter 6 is the conditional logit model, which deals with choice-specific characteristics (McFadden, 1974).

Explanatory variables in probability models fall into two categories. Models discussed so far use explanatory variables that are invariant to response categories. Demographic variables such as one's age, racial and ethnic background, and sex, and socioeconomic variables such as education, income, and occupation do not vary according to the response category chosen by the individual; they vary only across individuals. Another type of explanatory variables is choice specific; the variables take on different values dependent on the response category, even for the same individual.

In transportation research, where conditional logit was first applied, researchers study commuters' choice of modes of transportation to work, such as car, bus, and subway, while including alternative-specific attributes, which may include travel time, cost, or even perceived convenience. These independent variables take on values that are specific to each choice. Another example in transportation research is the choice of shopping centers. Consumers' travel time spent out of vehicle as well as in vehicle to a shopping center, travel cost, and attraction of a shopping center are all choice specific. Political scientists may study people's choice of presidential candidates by using alternative-specific attributes such as the respondent's perception of strength (or weakness) of a particular candidate in domestic versus foreign policy, or in finer detailed areas such as health care, education, national economy, national defense, and the like. Demographers may study individual's choice of contraceptives with respect to cost, convenience, and efficiency of a particular contraceptive choice. All these are examples of choice-specific characteristics or alternative-specific attributes (for more detailed discussion of types of explanatory variables, see Wrigley, 1985). A conditional logit model is called for when explanatory variables like these are present.

The Model

The conditional logit model estimates the effects of a set of choice-dependent variables on a dependent variable with unordered response categories. The equation,

$$\text{Prob}(y = j) = \frac{e^{\sum_{k=1}^{K} \alpha_k z_{jk}}}{\sum_{j=1}^{J} e^{\sum_{k=1}^{K} \alpha_k z_{jk}}}, \qquad [7.1]$$

gives $\text{Prob}(y = j)$ where $j = 1, 2, \ldots, J$. Note that the z variables have two subscripts in the model, k for distinguishing z variables, and j for distinguishing response categories, while parameters α do not change according to response category j. The reader will notice that if $J = 2$, Equation 7.1 simplifies to a binary version of the conditional logit model. A major difference between Equation 7.1 and an equivalent for the multinomial logit model (Equation 6.1) is the $\alpha_k z_{jk}$ term, which is a set of choice-dependent variables and their parameters. Another difference is that because α_k do not change across response categories, the exponentiation of $\sum_k \alpha_k z_{jk}$ will not be equal to 1 for $j = J$, as in the multinomial logit model.

Thus the "1 +" term drops out of the equation and the summation in the denominator changes from $j = 1$ to $J - 1$ in the multinomial logit model to $j = 1$ to J in the conditional logit model (also see Note 3).

We may, if we wish, express the conditional logit model in logit form. Equation 7.1 implies the following:

$$\log\left[\frac{\text{Prob}(y=j)}{\text{Prob}(y=J)}\right] = \sum_{k=1}^{K} \alpha_k(z_{jk} - z_{JK}). \qquad [7.2]$$

The steps leading from Equation 7.2 back to Equation 7.1 are identical to those for the multinomial model. When $J = 2$, Equation 7.2 simplifies to a binary version of the conditional logit model. The multinomial logit link function,

$$\eta_j = \log(\mu_j/\mu_J),$$

applies to the conditional logit model as well.

We may also specify a mixed model with both choice-specific and individual-specific attributes, thus combining the multinomial and the conditional logit model. Therefore, for modeling the probability $(y = j)$, we have

$$\text{Prob}(y=j) = \frac{e^{\sum_{k_1=1}^{K_1} \beta_{jk_1} x_{k_1} + \sum_{k_2=1}^{K_2} \alpha_{k_2} z_{jk_2}}}{\sum_{j=1}^{J} e^{\sum_{k_1=1}^{K_1} \beta_{jk_1} x_{k_1} + \sum_{k_2=1}^{K_2} \alpha_{k_2} z_{jk_2}}}, \qquad [7.3]$$

where both choice-specific explanatory variables, zs, and individual-specific variables, xs, are included in the same model, and the k_1 and k_2 subscripts separate the two types of explanatory variable. The mixed model will be useful for a variety of social science applications (for an exposition of a demographic example, see Hoffman & Duncan, 1988).

Typically, researchers include alternative-specific constants (ASCs) in a conditional logit model with choice-specific variables only by using a set of (1 to $J - 1$ number of) dummy variables representing the response categories to capture the mean effect of the unobserved factors in the error terms (Ben-Akiva & Lerman, 1985; Wrigley, 1985). This practice is not necessary when we have a truly mixed logit model, because the set of β_{jk}s already include ASCs (as in Equation 7.3). A conditional logit model with ASCs but not other x variables for the probability $(y = j)$ takes the form

$$\text{Prob}(y = j) = \frac{e^{\beta_{j_1} x_1 + \sum_{k=1}^{K} \alpha_k z_{jk}}}{\sum_{j=1}^{J} e^{\beta_{j_1} x_1 + \sum_{k=1}^{K} \alpha_k z_{jk}}}, \qquad [7.4]$$

where $x_1 = 1$ for all $J - 1$ categories, β_{j_1} are ASCs, and no subscript for k is needed anymore. The parameter β_{J_1} is typically normalized to zero. The conditional logit model with ASCs has been widely used in transportation research.

Conditional logit models can be estimated using the procedure PROC MLOGIT in SAS 5.18, which fits multinomial, conditional, or mixed logit models.[7] The discrete choice procedure in LIMDEP can also be used for estimation of this model. The requirement of the arrangement of the input data, however, differs between these two statistical software packages.

Due to similarities between the multinomial and the conditional logit models, the two models share similar advantages and shortcomings. For instance, conditional logit models are also subject to the problems related to the property of IIA. On the other hand, our ways of interpreting results from multinomial logit models can be adapted to interpret conditional logit models. A greater amount of care is necessary, however, in interpreting parameter estimates from a conditional logit model.

Interpretation of Conditional Logit Models

1. Marginal Effect on η or Transformed η. Unlike coefficients for individual-specific explanatory variables, those for alternative-specific attributes do not vary across response categories, as indicated by lacking the subscript j (see Equation 7.2). As a result, the interpretation of their effect on η or on transformed η is actually simplified. The effect of a choice-specific variable on the log-odds or the odds stays unchanged regardless of which two response categories are contrasted.

For an example using the conditional logit model with ASCs but no other x variables (Equation 7.4), let us look at commuting alternatives in an urban area in Tasmania, Australia, following the collapse of the Tasman Bridge.

TABLE 7.1

Sample Mean of Alternative-Specific Attributes,
Commuting Modes Alternative to the Tasman Bridge ($N = 1,324$)

Choice Alternative	Walk Time	Wait Time	In-Vehicle Time	In-Vehicle Cost	Out-of-Vehicle Cost
Car Driver (Bailey)	1.09	0	47.9	122.2	8.4
Car Passenger (Bailey)	0.88	0	51.9	7.1	0.2
Ferry Passenger	14.3	2.5	35.2	59.7	0.1
Car Driver (Punt)	0.3	0.4	66.6	134.7	2.4
Bus (Bailey)	15.2	2.0	31.9	55.0	0
Mean Total	10.39	1.78	39.2	73.4	2.16

SOURCE: Reprinted from *Transportation Research* 15B, D. A. Hensher, "A practical concern about the relevance of alternative-specific constants for new alternatives in simple logit models" pp. 407-410, Copyright © 1981, with kind permission from Elsevier Science Ltd, The Boulevard, Langford Lane, Kidlington OX5 1GB.
NOTE: All time is expressed in minutes; all cost is expressed in Australian cents. "Bailey" refers to commuting by the route of Bailey Bridge, a detour. "Punt" refers to commuting by way of the Risdon Punt, also a detour.

A full-loaded bulk ore carrier collided with the Tasman Bridge and severely damaged it on January 5, 1975. The bridge, built in 1964, spans the River Derwent and plays a vital role in facilitating commuters in the Hobart urban area. In 1974, prior to the accident, 28.2% of the population in the Hobart urban area lived on the east side of the river, an area providing only 5.3% of the employment. On a typical weekday in 1974, 43,930 vehicle crossings of the bridge were made. Alternative means of crossing were established following the collapse of the bridge and before its reopening more than 2½ years later. Two ferry routes for foot passengers only operated near the bridge; commuters could drive, either as a driver or as a passenger, several kilometers upstream to Bailey Bridge (which could take vehicles); they could use the Risdon Punt (which could take both vehicles and foot passengers), a bit closer than the Bailey Bridge; they could take a bus via the Bailey Bridge. Therefore, there were in total five commuting choices ($J = 5$). Data were collected on commuters in March 1977, regarding their mode of travel, route, walking time, waiting time, in-vehicle travel time, in-vehicle cost, out-of-vehicle travel time, and so on. For more detailed descriptions of the event and the survey, see Hensher (1979, 1981) or Wrigley (1985). The sample means of the five major alternative-specific explanatory variables are given in Table 7.1.

Although individual-specific variables may also have different mean values across response categories, they do not depend on the response

TABLE 7.2

Parameter Estimates for Alternative-Specific Attributes,
Commuting Modes Alternative to the Tasman Bridge ($N = 1{,}324$)

z & x Variable	$\hat{\alpha} \ or \ \hat{\beta}$	t Statistic
Walk Time	−0.05443	5.2
Wait Time	−0.12236	6.3
In-Vehicle Time	−0.01759	3.4
In-Vehicle Cost	−0.00736	5.2
Out-of-Vehicle Cost	−0.01810	3.0
ASC		
Car Driver (Bailey)	0.22930	0.7
Car Passenger (Bailey)	−0.64801	0.8
Ferry Passenger	1.81650	6.3
Car Driver (Punt)	−1.57860	3.5
Bus (Bailey)	—	—
LR Statistic	718.7	
df	9	

SOURCE: Reprinted from *Transportation Research* 15B, D. A. Hensher, "A practical concern about the relevance of alternative-specific constants for new alternatives in simple logit models" pp. 407-410, Copyright © 1981, with kind permission from Elsevier Science Ltd, The Boulevard, Langford Lane, Kidlington 0X5 1GB.

category. In contrast, the explanatory variables in Table 7.1 are choice-dependent. Their values vary according to the commuting alternative chosen. Despite their different choice-dependent values, these variables have only one set of α estimates. The coefficient estimates are presented in Table 7.2.

The parameter estimates for the explanatory variables all bear a negative sign, suggesting that the more time or cost involved, the less likely an alternative route is taken. Because of the property of the α's invariance to the choice category, the interpretation is more straightforward than with a multinomial logit model. the odds of choosing a particular commuting route with an additional minute in walking time are $\exp(-0.05443) = 0.94702$ times as high as choosing another route. Similarly, the odds of choosing a particular route with an additional minute in waiting time are $\exp(-0.12236) = 0.88483$ times as high as using another route. An increase in one cent of in-vehicle cost would make the odds of choosing an alternative decrease by a factor of $\exp(-0.00736) = 0.99267$. But that is the effect of one cent only. An increase in 25 cents of in-vehicle cost would make the odds of choosing an alternative decrease by a factor of $\exp(-0.00736 \cdot 25) = 0.83194$.

2. Predicted Probabilities Given a Set of Values in the Explanatory Variables. Using Equation 7.1 and statistics from Tables 7.1 and 7.2, we can calculate estimated probabilities for the five commuting alternatives in the example. To illustrate, let us examine how to calculate the probability for a car driver via the Bailey Bridge when all explanatory variables are kept at choice-specific mean levels. To simplify, the process is presented in two steps. The first step involves the calculation before exponentiating $\sum_k \alpha_k z_{jk}$ with $A = \sum_k \alpha_k z_{1k}$ for Bailey car drivers, $B = \sum_k \alpha_k z_{2k}$ for Bailey car passengers, $C = \sum_k \alpha_k z_{3k}$ for ferry passengers, $D = \sum_k \alpha_k z_{4k}$ for Punt car drivers, and $E = \sum_k \alpha_k z_{5k}$ for bus passengers:

$$A = 0.22930 \cdot 1 - 0.05443 \cdot 1.09 - 0.12236 \cdot 0 - 0.01759 \cdot 47.9$$

$$- 0.00736 \cdot 122.2 - 0.01810 \cdot 8.4$$

$$B = -0.64801 \cdot 1 - 0.05443 \cdot 0.88 - 0.12236 \cdot 0 - 0.01759 \cdot 51.9$$

$$- 0.00736 \cdot 7.1 - 0.01810 \cdot 0.2$$

$$C = 1.81650 \cdot 1 - 0.05443 \cdot 14.3 - 0.12236 \cdot 2.5 - 0.01759 \cdot 35.2$$

$$- 0.00736 \cdot 59.7 - 0.01810 \cdot 0.1$$

$$D = -1.57860 \cdot 1 - 0.05443 \cdot 0.3 - 0.12236 \cdot 0.4 - 0.01759 \cdot 66.6$$

$$- 0.00736 \cdot 134.7 - 0.01810 \cdot 2.4$$

$$E = - 0.05443 \cdot 15.2 - 0.12236 \cdot 2.0 - 0.01759 \cdot 31.9$$

$$- 0.00736 \cdot 55.0 - 0.01810 \cdot 0 . \qquad [7.5]$$

The second step uses the results from Step 1 and plugs them in Equation 7.4:

$$\text{Prob}(y = 1) = \frac{e^A}{e^A + e^B + e^C + e^D + e^E} = 0.1439 . \qquad [7.6]$$

This gives us the predicted probability for car drivers via the Bailey Bridge. By substituting e^B, e^C, e^D, or e^E into the numerator, the probability for the

TABLE 7.3

Predicted Probabilities for the Commuting Modes
Alternative to the Tasman Bridge

Choice Alternative	Walk Time-Low	Walk Time-Mean	Walk Time-High
Car Driver (Bailey)	0.1268 (0)	0.1439 (1.09)	0.1703 (2)
Car Passenger (Bailey)	0.1330 (0)	0.1527 (0.88)	0.1786 (2)
Ferry Passenger	0.6098 (10)	0.5811 (14.3)	0.5299 (20)
Car Driver (Punt)	0.0145 (0)	0.0172 (0.3)	0.0205 (1)
Bus (Bailey)	0.1159 (10)	0.1051 (15.2)	0.1006 (20)

NOTE: All probabilities are estimated with every choice-specific variable kept at its sample mean except Walk Time, whose values used in the calculation are given in parentheses.

other four alternatives will be derived. The five predicted probabilities are presented in the middle column of Table 7.3.

It is easy to see that on average the ferry routes were by far the most popular choice of commuting alternatives. Commuters were least likely to choose driving to Risdon and get the car on a punt to cross the river. They were about as likely to use the Bailey Bridge as a driver or a passenger. A little less popular was taking a bus via the Bailey Bridge. To aid interpretation and get a sense of the range of probabilities with respect to the range of a particular explanatory variable, I estimated the probabilities at low and high amounts of walking time for the five choices. The values are reasonably chosen at 0, 0, 10, 0, and 10 as the low-end values for the five choices, respectively; the high-end values are chosen as 2, 2, 20, 1, and 20 for the choices, respectively. The estimated probabilities for the low and the high amount of walking time are given in columns 1 and 3 in Table 7.3. They suggest the correspondence between the range of walking time for all the alternatives and the probability of choosing these alternatives.

There are almost an unlimited number of ways to interpret the results by using predicted probabilities in conditional logit models. Instead of changing the values of an explanatory variable for all choices, one may choose to change the level of that variable for one choice only, two choices only, and so on. If one changes the value of an explanatory variable for one choice only, the direction of change in predicted probabilities will be predictable: The predicted probability for the alternative with which the change in the explanatory variable is associated will change in one direction and all other predicted probabilities for the other response categories will change in the opposite direction. In addition, one may want to change the value of more than one choice-specific explanatory variable. In that event, more than one table will be needed for presentation. One can also use a finer gradation between the low and the high values of an explanatory variable, and present the results in graphs.

3. Marginal Effect on the Probability of an Event. As with predicted probabilities, alternative-specific explanatory variables make the interpretation using marginal effects on response probability more complex. Because explanatory variables are choice-specific, the marginal effect due to the change in the level of a variable related to one response category in principle will be different from the effect due to the change in the level of the same variable related to another response category. Thus marginal effects in conditional logit models, though similar to those in multinomial logit models, need more care when interpreting. In a conditional logit model the z_{jk} are choice-specific while the α_k are not. Unlike the effect of α on the odds ratio, the marginal effects do vary according to the response category in the dependent variable. This is so because of the choice-specific nature of the variables, of which the partial derivative is a function.

By differentiating Equation 7.4 for response probability with respect to z_{jk}, we find that the marginal effects of the regressors on response probability are

$$\frac{\partial P_j}{\partial z_j} = P_j(1 - P_j)\alpha \,,$$

$$\frac{\partial P_j}{\partial z_{j^*}} = P_j P_{j^*}\alpha \,,$$

[7.7]

where the subscript j, as before, refers to the response category to which a probability or a level in an alternative-specific explanatory variable is related and for which a marginal effect *is* being calculated, and the subscript j^* refers to the response category to which a probability or a level in an explanatory variable is related but for which a marginal effect is *not* being calculated.

To illustrate, let us examine the marginal effects of walking time on response probability with the explanatory variables held at their mean levels. Therefore, we can use the results in the middle column in Table 7.3. These five predicted probabilities, together with the α estimate for walking time, -0.05443, are used to estimate the marginal effects (Table 7.4).

TABLE 7.4
Marginal Effect of Walking Time on Choice Probability

Choice Alternative	Car Driver (Bailey)	Attribute Level of Walk Time for: Car Passenger (Bailey)	Ferry Passenger	Car Driver (Punt)	Bus (Bailey)
Car Driver (Bailey)	−0.0067	0.0012	0.0046	0.0001	0.0008
Car Passenger (Bailey)	0.0012	−0.0070	0.0048	0.0001	0.0009
Ferry Passenger	0.0046	0.0048	−0.0132	0.0005	0.0033
Car Driver (Punt)	0.0001	0.0001	0.0005	−0.0009	0.0001
Bus (Bailey)	0.0008	0.0009	0.0033	0.0001	−0.0051

NOTE: The marginal effects across response categories in a column or across alternative-specific levels of "Walk Time" in a row may not sum out to zero due to rounding error. The marginal effects, like the explanatory variables, are also choice-specific.

To calculate the marginal effects on the diagonal ($j = j^*$), the first line in Equation 7.7 is used. That is, $(0.1439)(1 - 0.1439)(-0.05443) = -0.0067$, $(0.1527)(1 - 0.1527)(-0.05443) = -0.0070$, $(0.5811)(1 - 0.5811)(-0.05443) = -0.0132$, and so on. For calculating the marginal effects in the off-diagonal cells ($j \neq j^*$), the second line in Equation 7.7 is used: For instance, for the cell in row 3 and column 1, $(-0.5811)(0.1439)(-0.05443) = 0.0046$; for the cell in row 3 and column 2, $(-0.5811)(0.1527)(-0.05443) = 0.0048$; for the cell in row 4 and column 3, $(-0.0172)(0.5811)(-0.05443) = 0.0005$; for the cell in row 5 and column 3, $(-0.1051)(0.5811)(-0.05443) = 0.0033$; and so on. As with the calculation of predicted probabilities and marginal effects in earlier chapters, a spreadsheet software will facilitate the computation tremendously.

The 5-by-5 matrix of marginal effects is just for the effect of one alternative-specific variable, walk time. To present the marginal effects of the other variables, we will need additional panels in the table or, preferably, additional tables. These marginal effects are in fact quite intuitively appealing. For an additional minute of walking time for a car driver via the Bailey Bridge, the corresponding probability for selecting this route would decrease by approximately 0.0067, while the probabilities for the other four choices would increase, because this person might use one of the remaining routes. The probability for a car passenger via Bailey would increase by approximately 0.0012, that for a ferry passenger would increase by approximately 0.0046, that for a car driver via Risdon would increase by approximately 0.0001, and that for a bus passenger via Bailey would increase by 0.0008 approximately. As with any multinomial logit models, the marginal effects across the response categories should cancel out, as they indeed do for the marginal effect of walking time related to car drivers via the Bailey Bridge.

An additional minute in walking time spent by ferry passengers would decrease the probability for taking that route by about 0.0132, the greatest marginal effect in the whole table. One may attribute the size of the effect to the longer walking time of ferry passengers. Table 7.1 tells us that the mean walk time of bus passengers is actually about one minute longer than that of ferry passengers, yet an additional minute in walking time spent by bus passengers would decrease the probability for choosing the bus alternative by only about 0.0051. Again, other routes would appear relatively more attractive and make up for the loss in popularity of the ferry route due to increased walking time. Once more, the total marginal effects equal to zero. These effects on probability seem small, but remember that they are due to just one additional minute of walking time. Therefore, commuters were sensitive to how much time they would spend on walking.

A property not shared by marginal effects in multinomial logit models discussed in Chapter 6 is that the marginal effects in conditional logit models also cancel out one another across attribute levels of an explanatory variable related to the response categories. Try adding up the effects in a row across columns to find out. This feature suggests that the effect on a response category resulting from a change in the level of an explanatory variable specific to the same response category must be in balance with the effects on the response category resulting from a change in the level of the variable specific to the other response categories. This fact is implied by the α_k parameters, which are not choice-specific, indicating that an attribute such as walking time holds the same intrinsic value for different discrete choices. In multinomial logit models, the only way an explanatory variable can influence response probabilities is by having a differential impact represented by β_{jk}. In contrast, in conditional logit models the alternative-specific explanatory variables, z_{jk}, affect response probabilities, thus requiring only one set of parameters α_k.

A variation of this interpretation with marginal effects on probabilities is to examine the elasticities of probabilities. Although it is purely a matter of taste which one to report, some readers may find elasticities a bit too involved to understand because things are stated in relative terms. Interested readers are referred to Ben-Akiva and Lerman (1985) and Greene (1990).

8. POISSON REGRESSION MODELS

Sometimes we have an event count as a dependent variable that appears to be continuous. It is often mistakenly modeled by using a continuous variable approach such as linear multiple regression. Examples of such event counts include the number of visits to one's dentist or physician per year, daily crime counts in an urban area, the number of certain political events such as presidential vetoes in a given time interval, the number of certain international events in a given time period, and the number of newly created social organizations. All these events, measured by a nonnegative integer, are relatively rare and are assumed to be generated by a Poisson process. For modeling such data a Poisson regression model is required.

In recent years a number of interesting applications using the Poisson regression have appeared, following the classic example of the Poisson regression model of cargo ships damaged by waves (see McCullagh & Nelder, 1989). A Poisson regression model has been used to study U.S. presidential appointments to the Supreme Court (King, 1987), political party switching among the members of the U.S. House of Representatives (King, 1988), international wars (King, 1989a), daily homicide counts in California (Grogger, 1990), and foundings of day-care centers in Toronto, Canada (Baum & Oliver, 1992). This list is by no means complete.

The Model

Rare event counts are generated by a Poisson process and can be described by a Poisson distribution. The basic model is the single-parameter Poisson probability density function,

$$f(y_i, \theta_i) = P(Y_i = y_i) = \frac{e^{-\theta_i} \theta_i^{y_i}}{y_i!}, \quad \text{for } y_i = 0, 1, 2, \ldots, \infty; \ \theta_i > 0, \quad [8.1]$$

where θ_i is the expected value of Y_i, $E(Y_i)$. The equation above defines the model in terms of probability. We can express the model in terms of the observed y_i or its expectations θ_i. Then we have a common formulation of the Poisson regression

$$\ln \theta_i = \sum_{k=1}^{K} \beta_k x_{ik}. \quad [8.2]$$

This ensures that θ_i is always greater than 0 because $\theta_i = \exp(\sum_k \beta_k x_{ik})$. Equation 8.2 reminds us of the Poisson distribution-based logarithm link function in Chapter 2,

$$\eta = \log\mu .$$

Most of the time we cannot assume that populations at risk or observation intervals are constant. Obviously, even rare events will become more numerous if we observe long enough. Then we need to include a fixed variable, n, to reflect the amount of exposure to the event in the specification of the Poisson regression model:

$$\ln\theta = \ln n + \sum_{k=1}^{K} \beta_k x_k ,$$

$$\ln\frac{\theta}{n} = \sum_{k=1}^{K} \beta_k x_k .$$

[8.3]

Obviously this resembles the binary logit model. In this case we can actually use a log-linear-type model to estimate (see Fienberg, 1980, for the relation between log-liner models and Poisson distributions). Practically, make sure that your exposure variable is at least 10 to 100 times greater than the event count variable so that a Poisson distribution assumption may be appropriate. Otherwise you will not have a Poisson distribution, though you still can use a log-linear model for the analysis. Poisson regression models can be estimated with statistical software packages such as LIMDEP or GLIM. If we use the specification of Equation 8.3, some procedures for logit models can also be used to estimate a Poisson regression model. Specifically, the procedure PROC LOGISTIC in SAS is particularly easy to use, because it allows the inclusion of a population-at-risk or exposure variable such as n and automatically constrains its parameter to 1. Freeing the constraint up does no harm (Maddala, 1983), and sometimes makes substantive appeal if it is freed up and included as another explanatory variable (King, 1988). Although the SAS procedure estimates the Poisson regression model with the coefficient for the exposure variable fixed at unity, the Poisson regression procedure in LIMDEP has the flexibility to estimate the exposure variable with a free parameter.

The basic Poisson regression model assumes that the mean of y_i equals its variance, as does its namesake distribution. This imposition sometimes is not realistic. If the condition is not satisfied, such as when overdispersed data are fitted to the Poisson model, the estimated covariance matrix of the regression parameters is biased downward, resulting in overstated significance levels. In this event, rather than $\text{var}(y) = \theta$, we have $\text{var}(y) = \sigma^2\theta$, where σ^2 is the dispersion parameter. The dispersion parameter σ^2 can, if required, be estimated, as suggested by McCullagh and Nelder (1989), by

$$\hat{\sigma}^2 = \chi^2/(N-K) = \sum_{i=1}^{N} \frac{(y_i - \theta_i)^2}{\theta_i}/(N-K) . \qquad [8.4]$$

For large samples the denominator can be considered to be N.

There are a few ways to model overdispersed event counts. A common practice is to use the negative binomial regression model. For extended discussions of why the Poisson regression would fail and examples of applying the negative binomial model, see Grogger (1990) and King (1989a). King (1989c) proposes a generalized estimator that models overdispersed, Poisson-distributed, and underdispersed data. King (1989b) considers a seemingly unrelated Poisson regression model with two response variables and two correlated error terms as another extension to the regular Poisson regression.

Interpretation of Poisson Regression Models

1. Marginal Effect on η *or Transformed* η. In Poisson regression models the interpretation of marginal effect on $\exp(\eta)$ is quite straightforward and easy to use, because $\exp(\eta)$ is θ, the expected value of y. This relation is implied in Equation 8.2. Thus the marginal effect of x_k on expected y is given by $\theta\beta_k$. Alternatively, we may interpret the multiplicative effect of x_k on expected y by computing $\exp(\beta_k)$, similar to the interpretation of the effect on an odds.

Because θ is a function of all the x variables, we have two basic ways to interpret marginal effect of x_k on the expected value of y. We may use the sample mean of y, which mathematically is not equivalent to, but practically can be substituted by, having all the x variables held at their mean levels. Alternatively, we may specify a set of x values, get the corresponding θ, and interpret marginal effect accordingly.

For an empirical example, we examine the number of appointments to the U.S. Supreme Court in a given year as a Poisson process influenced by several explanatory variables (King, 1987). These variables are the number of appointments in the previous 6 years, the proportional change in the percentage of population who are military personnel on active duty, the percentage of members in the U.S. House of Representatives who were newly elected in the most recent election, and a squared term of the new House members variable. The natural-logarithm transformed variable of the number of seats in the Supreme Court is also included as the exposure variable. The substantive rationales for the inclusion of these variables are as follows (King, 1987): First, the number of seats on the Court has varied from 5 to 10 historically, making the expectation of the same number of retirements implausible. Second, a larger number of recent retirements may make the expected number of current appointments smaller. Finally, the justices may be hypothesized to have higher probabilities of retirement in times of political turmoil and realignment because individual justices do not change their attitudes very often or very quickly. Military conflict, as indicated by the proportional change in the percentage of population who are military personnel on active duty, signifies political, social, and economic turmoil. The extent of the electoral change, as indicated by the percentage of members in the U.S. House of Representatives who were newly elected in the most recent election, also implies political turmoil and realignment. Table 8.1 presents the parameter estimates from a Poisson regression model for this example.

The ratio of the estimates to their standard errors and the related p values for making a Type I error suggest that none of the variables can be ignored in studying U.S. Supreme Court appointments. Let us look at the effect of two variables, previous appointments and percentage military increase. Common sense suggests that numerous previous appointments should discourage making new appointments. This is confirmed by the estimate, -0.2184. As a condition for interpretation, the mean of the number of appointments in any year is 0.5131.[8] The effect of the number of appointments in the previous 6 years is $(0.5131)(-0.2184) = -0.1121$, suggesting that an additional appointment in the previous 6 years will decrease the expected number of appointments in the current year by approximately 0.1121, other things being equal. In other words, the difference between making no appointment and five appointments in the previous 6 years is more than half a person. The effect of percentage military increase, on the other hand, seems moderate. A 1% increase in the percentage of population who are military personnel will increase the expected number of appointments

TABLE 8.1

Parameter Estimates From a Poisson Regression Model
of U.S. Supreme Court Appointments, 1790-1984

x Variable	$\hat{\beta}$	se($\hat{\beta}$)	p
ln(No. of Seats)	1.7360	1.0120	0.0431
Appointments in Last 6 yrs	−0.2184	0.0715	0.0011
% Military Increase	0.4626	0.2258	0.0202
% New House Members	5.9000	4.6450	0.1020
(% New House Members)^2	−10.4200	6.5630	0.0562
Constant	−4.3540	2.4770	0.0394
LR Statistic		18.5	
df		5	

SOURCE: King (1987, Table 1).
NOTE: The probability that the coefficients for both percentage new House member variables in the
quadratic function are zero is less than 0.05. The parameter estimates for military increase, new House
members, and new House members squared are rescaled to reflect proportion instead of percentage in
the variables for facilitating presentation.

by approximately 0.0024, other things being equal, because (0.5131) ×
(0.004626) = 0.0024 (the estimate for military increase is divided by 100
because the estimate in Table 8.2 gives proportional change). But this
reflects only a 1% increase. In certain historical times the increase has been
much greater than that.

We can also specify a set of values of xs and examine the effect of x_k.
Under the circumstances of six judges serving in the Court, a mean number
of previous appointments (3.0785), no military increase, and 10% new
House members in the most recent election, the expected value of the
number of appointments is exp(−4.354 + 1.736 · ln(6) + . . . + 0.059 · 10 −
0.001042 · 10^2) = 0.2393 (again, the estimates for those percentage measures are divided by 100). The effect of appointments in the previous 6 years
then is (0.2393)(−0.2184) = −0.0523, and the effect of percentage military
increase is (0.2393)(0.004626) = 0.0011. These two have a much smaller
impact than the same effects calculated at the sample mean of y.

Alternatively, we may examine the multiplicative effect of x_k. In the
current example, the marginal effect of previous appointments becomes exp(−
0.2184) = 0.8038, suggesting that an additional appointment in the previous
six years will reduce the expected number of appointments by a factor of
0.8038, other things being equal. A 1% increase in the percentage of population
who are military personnel will increase the expected number of appointments

TABLE 8.2

Predicted Probability and Marginal Effect on Probability
From the Poisson Regression Model in Table 8.1

Condition	$y = 0$	$y = 1$	$y = 2$	$y = 3$	$y = 4$	$y = 5$
Predicted Probability						
New House Members						
= 10%	0.78718	0.18837	0.02254	0.00180	0.00011	0.00001
At Sample Mean of y	0.59864	0.30716	0.07880	0.01348	0.00173	0.00018
Marginal Effect of % Military Increase on Probability						
New House Members						
= 10%	−0.00087	0.00066	0.00018	0.00002	0.00000	0.00000
At Sample Mean of y	−0.00142	0.00069	0.00054	0.00016	0.00003	0.00000

NOTE: For the condition of new House members = 10%, number of seats is held at six, appointments in the previous 6 years at the mean value from 1790 to 1980 (Ulmer, 1982), and percentage military increase at zero. For the sample mean of y, the value for 1790 to 1980 (Ulmer, 1982) is used. Predicted probabilities may not sum up to 1 due to rounding error.

by 1.0046 times, other things being equal, because exp(0.004626) = 1.0046. The advantages of the multiplicative effect are that it is not affected by the values of y and that it is a bit more accurate than the partial derivative because the latter gives an approximation of marginal effects. Its major drawback is that it appeals less intuitively to some researchers because the marginal effect gives an expected increase or decrease by the number of event counts while the multiplicative effect gives the expected increase or decrease by the number of times. The reader is encouraged to compare the two ways for the effect of both dichotomous and continuous variables by using the additive effect to simulate the multiplicative effect.

2. Predicted y Given a Set of Values in the Explanatory Variables. There are two types of predicted values worth interpreting in Poisson regression models: predicted y and predicted probability that $Y = y$. Let us examine them separately. From Equation 8.2 we know that the expected value of y is $\exp(\sum_k \beta_k x_k)$. This gives us a way for calculating predicted y almost as easy as in classical linear regression models. We may, for instance, want to predict a sequence of y by changing the values in one (or two) variables while holding all others at their means or certain other levels.

Continuing the U.S. Supreme Court appointment example, let us calculate expected y by changing the percentage of new House members from 5 to 30 with the increment of 5 while holding the number of seats at six, the number of appointments in previous years at their historical mean of 3.0785, and the percentage increase in military personnel at zero. Therefore, for the six levels of new House member percentages, we have the predicted number of appointments as follows:

% New House Members	5	10	15	20	25	30
Predicted Count	0.192	0.239	0.282	0.316	0.336	0.338

Thus we can easily see how expected event count varies with the level of an independent variable. In the current example, the predicted number of appointments increases at a faster pace at lower levels of percentage new House members, and the increase levels off when the percentage of new House members is high because of the quadratic function.

3. Predicted Probabilities Given a Set of Values in the Explanatory Variables. The second type of predicted value in a Poisson regression model is probability, because Poisson regression is, after all, a probability model. Using Equation 8.1, we can calculate the predicted probability that the event count equals 0, 1, 2, 3, and so forth, given certain x values.

For the current example of U.S. Supreme Court appointments, let us use the same set of given values in the x variables from the previous subsection and 10% of new House members, get the result equivalent to the expected y given above, and plug it in Equation 8.1 as θ. To facilitate the reader with the calculation without necessarily giving too much detail, I only present the steps for calculating predicted Prob($Y = 3$):

$$\text{Prob}(Y = 3) = \frac{e^{-0.2393} \cdot 0.2393^3}{1 \cdot 2 \cdot 3} = \frac{0.0108}{6} = 0.0018 . \qquad [8.5]$$

The other probabilities, which can be calculated accordingly, are presented in the first row in Table 8.2. Therefore, for a fixed level in an independent variable, we will have a schedule of probabilities. This means that for every level in an x we will have a separate schedule of predicted probabilities.

The schedule of probabilities for 10% of new House members shows a typical Poisson distribution concentrated at low levels of event counts. Alternatively, we may predict these probabilities at the sample mean value of y—results reported in the second row of Table 8.2. Using the sample mean, there are more appointments predicted than when using the first set of conditions with 10% of new House members. As with predicted probabilities for other probability models, the sum of all probabilities should be equal to unity. With a Poisson distribution, it is impossible and unnecessary to estimate a probability for all response categories because the probability for all categories beyond a certain response level are approximately zero. The predicted probabilities in either row in the table add up to approximately one. Another way to check the result when predicting at the sample mean level is to see if $\sum_j y \cdot \text{Prob}(Y = y) = \theta$. We have: $0 \cdot 0.59864 + 1 \cdot 0.30716 + 2 \cdot 0.07880 + 3 \cdot 0.01348 + 4 \cdot 0.00173 + 5 \cdot 0.00018 = 0.51302$. This is approximately the same as θ.

4. Marginal Effect on the Probability of an Event. Because Poisson distributions are discrete, it still makes sense to use both the predicted probability and the marginal effect on response probability. The marginal effect on probability in the Poisson regression model tells the expected probability that $Y = y$ (0, 1, 2, 3, etc.) given a unit change in x_k.

Taking the partial derivative of $\text{Prob}(Y = y)$ with respect to x_k, we have

$$\frac{\partial \text{Prob}(Y = y)}{\partial x_k} = \frac{\beta_k \theta e^{-\theta}(y\theta^{y-1} - \theta^y)}{y!}, \qquad [8.6]$$

where $y = 0, 1, 2, 3$, etc.[9] To illustrate, we continue with the example of U.S. Supreme Court appointments. As before, we use the set of given values of six seats in the court, average appointments in the previous 6 years as 3.0785, no military increase, and 10% new House members. To interpret the effect of percentage military increase on response probability, we take the expected event count value ($\theta = 0.2393$) from a previous subsection, and plug it and percentage military increase ($\beta_3 = 0.0046$) in Equation 8.6. The marginal effect on the probability of zero appointment is then

$$\frac{\partial \text{Prob}(Y = 0)}{\partial x_3} = \frac{0.0046 \cdot 0.2393 e^{-0.2393}(0 \cdot 0.2393^{0-1} - 0.2393^0)}{0!}$$

$$= \frac{0.0046 \cdot 0.2393 \cdot 0.7872(-1)}{1} = -0.00087. \qquad [8.7]$$

This gives the first entry of the results in the third row in Table 8.2. As with the predicted probability, the marginal effect gets smaller as y gets larger. The marginal effect of percentage increase in the percentage of population that are military personnel on response probability is also estimated when the sample mean θ is used (the last row in Table 8.2). Thus with 1% increase in the proportion of the U.S. population that are military, the probability that no appointment to the Supreme Court is made in a given year will decrease by approximately 0.00142, the probability that one appointment is made will increase by approximately 0.00069, the probability that two appointments are made will increase by approximately 0.00054, and so on.

As with marginal effects in other probability models, the estimated marginal effects in Poisson regression should also cancel themselves out across the response categories. Because the effect is already negligible when $y = 5$, the marginal effects indeed sum out to zero in both rows. Similar to the marginal effects on probability in ordinal logit or probit models, the change in x_k makes some response categories reduce, and others enlarge, their probabilities. If β_k is positive, the effect should be positive for greater y and negative for smaller y. If β_k is negative, the effect should be negative for greater y and positive for smaller y. This again serves as a quick check on the correctness of the calculation.

The same results for the marginal effect can be derived by taking the difference of the corresponding predicted probabilities. For instance, if we use the predicted probability for $y = 2$ when new House members are 10% in row 1 of Table 8.2 and calculate the same probability by changing the value of percentage military increase to 1, the difference of these two probabilities will yield 0.00018. Consistency between the two methods is very high for a variable such as percentage military increase. The cruder the measurement (in terms of fewer measurable levels), the less consistent the two methods become. The extreme case is that of a dummy variable, for which the marginal effect calculated using Equation 8.6, although in the neighborhood of the correct estimate, differs by a much larger margin from the effect estimated by taking the difference of two probabilities. As shown in Chapters 3, 4, and 5, this comment about calculating marginal effects of dummy variables applies to all probability models.

As with predicted probabilities, marginal effects on response probability can be estimated at a variety of values in xs and presented accordingly. Graphics are quite useful for presenting marginal effects at two or more sets of different x values and when interpreting with predicted probabilities.

Poisson regression models are a useful statistical method for analyzing count data. Often these data are wrongly studied by using OLS regression models (see King, 1988). Understanding how to interpret results from a Poisson regression model will help the reader feel more comfortable with using such a model. As demonstrated, Poisson regression models have four ways of interpretation, which are reasonably straightforward to apply in social science research.

9. CONCLUSION

I have three goals in the concluding chapter. First, I summarize the ways of interpreting parameter estimates from a variety of probability models. Second, I briefly discuss some texts more formally introducing and discussing these models. Finally, I comment on some important implications of interpreting probability models from the perspective of generalized linear models.

Summary

In Chapter 2, I introduced the generalized linear model and four systematic ways of interpreting parameter estimates from a generalized linear model—predicted values of η or transformed η given a set of values in the explanatory variables, marginal effect on η or transformed η, predicted probabilities given a set of values in the explanatory variables, and marginal effect on the probability of an event. The common practice of examining the sign and the significance of parameter estimates, though easy to use and thus unnecessary to discuss in this volume, does not utilize the rich information these probability models provide. Because the probability models discussed in the book are all special cases of the generalized linear model with their link functions determined by the underlying distribution of the data and our theoretical understanding of the distribution, the four ways of interpretation are readily extended to interpreting results from all of these probability models. Therefore, the systematic ways of interpreting probability models allow us better substantive understanding of the results.

I then examined with examples how to use the three (or four) ways of interpretation in a variety of probability models—binary logit and probit models in Chapter 3, sequential logit and probit models in Chapter 4, ordinal logit and probit models in Chapter 5, multinomial logit models in Chapter 6, conditional logit models in Chapter 7, and Poisson regression models in Chapter 8. The four ways of interpretation—predicted values of η or transformed η given a set of values in the explanatory variables, marginal effect of x_k on η or transformed η, predicted probabilities given a set of values in the x variables, and marginal effect of x_k on response probability—have the same manner of interpretation in all the models examined but different formulae for calculation because of the link function unique to each probability model. The first of the four ways, however, is not as useful except for the Poisson regression and possibly the logit model.

All the probability models examined can be written in two general forms of the dependent variable—probability and a quantity based on observed count, η. In principle we may interpret parameter estimates via marginal effect and predicted values for either probabilities or quantities. Thus we could have four ways of interpretation. For most of these probability models, however, predicted values of η are not very meaningful because η is often a logit or probit, which does not lend itself to easy interpretation. Unlike most other models, the Poisson regression has event count as a transformed η, which makes much intuitive and practical sense.

Major Texts of Probability Models

Most monographs introduced in Chapter 1 treat probability models to a varying degree of extensiveness and difficulty. In the following I review some of the major ones to highlight their major features and differences. Because many of these books cover vitally important topics such as model specification (in much greater detail) and estimation, readers unfamiliar with these issues may consult some of these works.

First, because my ways of interpretation are discussed from the perspective of generalized linear models, some readers may want to study the topic more formally and systematically. McCullagh and Nelder (1989) give probably the most authoritative treatment of the generalized linear model. For an easier introduction, and a thinner volume, the reader is referred to Dobson (1990). Either of these works should give the reader enough exposure to the topic.

In addition to the paradigm of the generalized linear model, two traditions prevail for presenting some (or all) of these probability models. They are (1) the log-linear model approach and (2) the regression model approach in which the dependent variable is qualitative, discrete, or simply limited. Almost all the contents in these two traditions can be subsumed under the overarching generalized linear model (see McCullagh & Nelder, 1989).

Many general treatments of log-linear models have become available over the past decade or so. They include Agresti (1984, 1990), Bishop et al. (1975), Fienberg (1980), and Knoke and Burke (1980). An interesting type of log-linear model, the association model, also has received much attention [see Clogg (1982, and his many other articles); Goodman (1991, and his many other articles)]. More closely related to the type of probability model discussed here, however, are two popular texts by Agresti (1984, 1990). Although starting from the foundation of contingency tables and log-linear models, Agresti (1990) attempts to integrate generalized linear models as well. This shows that the line between the first and second tradition has become blurred. A recent Sage QASS publication by DeMaris (1992), a clear presentation of topics from odds ratios in contingency tables to ordinal logit models, though based on the first tradition, attempts to bridge the log-linear-model approach for contingency tables with the logit-model approach for individual-level data.

Another Sage QASS publication, the well-received number by Aldrich and Nelson (1984), belongs in the second tradition. This text focuses on the comparison of linear, probit, and logit models in modeling response probability and why OLS linear models would fail when the response variable is dichotomous. The coverage extends to multinomial logit models. Next we have several texts treating the topic of the analysis of discrete data. Cox and Snell (1989) discuss specifically the analysis of binary response data. For the analysis of discrete data in general, four major texts come to mind. They are Ben-Akiva and Lerman (1985), Santer and Duffy (1989), Train (1986), and Wrigley (1985). Ben-Akiva and Lerman's (1985), Train's (1986), and Wrigley's (1985) are written for a targeted readership— the first two for transportation researchers and the last for geographers and environmental scientists, with respective examples included. The organization of Train's (1986) book reveals a major objective of his: The second half of the book is devoted to an application of qualitative choice models to automobile demand. All three texts, however, are written at a level that is accessible to many social scientists at large. These books deal with a variety of discrete choice models including multinomial logit and probit, and conditional logit models. Also valuable in them is the discussion of the IIA property. For a more theoretical and mathematical treatment of probability models for discrete data, Santer and Duffy (1989) is a good choice.

Much of the regression model tradition is closely related to econometrics. Thus it is not surprising that econometricians contribute heavily to this tradition of research. Emerging from the more technical literature are several comprehensive treatments written at a level still accessible to quite a few social scientists. Monographs worth mentioning include Greene (1990; or the 1993 second edition) and Maddala (1983). Maddala's well-organized and systematic discussion of limited-dependent variables covers many probability models and other models dealing with censored dependent variables and sample selection bias and has been considered a highly important contribution to the topic. Greene's (1990) recent textbook, though written as a general econometric text, covers tobit and sample selection models (also models of duration data in the 1993 second edition) as well as all the probability models (except sequential models) discussed here. The discussion of these models is lucid yet has enough technical detail to explain maximum likelihood estimation of these models. Another strength of the book is the inclusion of materials on the principle of maximum likelihood estimation and optimization procedures. Some econometric review articles covering logit, probit, and related models should also be mentioned; they include Amemiya (1981) and McFadden (1976, 1982).

For readers who would like to brush up on the foundation that underlies all these probability models—the classical linear regression model—many entries in the Sage QASS series will be excellent choices for understanding such linear models and their interpretation. They include Achen (1982), Berry and Feldman (1985), Hardy (1993), Jaccard, Turrisi, and Wan (1990), Lewis-Beck (1980), and Schroeder, Sjoquist, and Stephan (1986). In addition, a couple of the entries deal with assumptions (Berry, 1993) and diagnostics (Fox, 1991). While there are texts with more extensive treatment of linear regression models, the above Sage publications taken together do an excellent job of covering many topics related to linear regression models.

Further Comments on Interpreting Probability Models

The interpretation of parameter estimates from probability models is by no means confined to the four general ways presented in this book. The four ways may readily extend to additional transformed functions for interpretation. For example, marginal effect on response probability is defined by the partial derivative of response probability with respect to x_k. By taking the partial derivative of the natural-logarithm transformed probability with respect to the natural-logarithm transformed x_k, we have the elasticity interpretation, common in economics. For multinomial logit models such as the conditional logit model, the elasticity interpretation may be especially useful (see Ben-Akiva & Lerman, 1985; Greene, 1990). Even though it may be useful, it is also a bit more involved to use. Interested readers should consult the references above.

Another type of statistical model resembles the family of probability models examined in this book. The hazard rate models in event history or survival analysis are also widely used in social science applications. These include both parametric models such as the exponential, the Gompertz, and the Weibull model, and semiparametric models such as Cox's proportional hazards models (see, e.g., Allison, 1984; Lawless, 1982). Although these models for survival data can also be subsumed under the generalized linear model, their specifications differ more from the probability models we examined than these probability models do among themselves. With discrete-time data, the hazard rate model can be approximated by a logit model. Interpretation of results from such a model can follow the four ways of interpretation. In general, however, interpretation of results from hazard rate models has some unique properties, and thus is beyond the scope of the present book.

These four ways of interpreting results from probability models are not without shortcomings. Specifically, predicted probabilities are merely probabilities estimated with the sample data points. What about their confidence intervals? Although we may get some general idea from the likelihood ratio statistic for the overall model and the standard errors of the parameter estimates, the width of the confidence interval is unclear. When making predictions in linear regression models, we can construct confidence intervals to assess how precise the predictions are. Similarly, we need to assess how precise the predicted probabilities are. The necessity of considering confidence intervals of predicted probabilities increases with the closeness between predicted values of response categories. We can construct confidence intervals for predicted probabilities from these probability models, which, as does the classical linear regression model, all belong to the family of generalized linear models (Fox, 1987; Liao, 1993). Such confidence intervals, a topic beyond the scope of this monograph, will prove useful in many social science applications.

NOTES

1. More generally, both the logit link and the multinomial logit link are special cases of the multivariate logit link McCullagh and Nelder (1989) discussed.

2. Without having the original data to fit a logit model, the logit estimate is derived by multiplying the probit estimate by a factor of 1.6, as proposed by Amemiya (1981).

3. The other common version of the same formulation is written as

$$\text{Prob}(y=j) = \frac{e^{\sum_{k=1}^{K} \beta_{jk}x_k}}{\sum_{j=1}^{J} e^{\sum_{k=1}^{K} \beta_{jk}x_k}}, \qquad [A.1]$$

where $j = 1, 2, \ldots, J$. The parameters β_J are normalized to be zero. This version is more compact because it implies both Equations 6.1 and 6.2. Equations 6.1 and 6.2 are presented in the text because they will facilitate interpretation more readily for most readers than the single equation above.

4. For a response variable with J categories, there are $J - 1$ nonredundant sets of parameter estimates only. For J response categories, however, there exist $J!/2!(J - 2)!$ number of contrasts between the categories.

5. The term $\beta_2 + 2\beta_3 x_2$ is just the partial derivative of the log-odds with respect to x_2 for $\log[\text{prob}(y = j)/\text{prob}(y = J)] = \beta_1 + \beta_2 x_2 + \beta_3 x_2^2 + \ldots + \beta_K x_K$.

6. These parameter estimates correspond to the contrast of "contraceptive reason" versus "mixed reasons," of "no sterilization" versus "mixed reasons," and of "medical reason" versus "mixed reasons." They are from the rows related to parity in the first three columns of Table 6.1: 1.05, −0.29; −1.04, 0.16; and 0.13, 0.02.

7. SAS Institute may not support this procedure in version 6 and later of the software once it has completed its upgrade from FORTRAN by rewriting the software in C, because the MLOGIT procedure is user-written and user-supported.

8. The mean of y is not reported in King (1987). Thus an approximation is used by the mean of number of U.S. Supreme Court Appointments for the period 1790-1980, 98/191 = 0.5131, from Ulmer (1982).

9. The derivation of Equation 8.6 is as follows:

$$\frac{\partial \text{Prob}(Y = y)}{\partial x_k} = \frac{\frac{\partial(e^{-\theta}\theta^y)}{\partial x_k}}{y!}$$

$$= \frac{\theta^y \frac{\partial}{\partial \theta}\frac{\partial \theta}{\partial x_k} + e^{-\theta}\frac{\partial}{\partial \theta}\frac{\partial}{\partial x_k}}{y!}$$

$$= \frac{\theta^y e^{-\theta}e^{\sum_{k=1}^{K}\beta_k x_k}\beta_k + e^{-\theta}y\theta^{y-1}e^{\sum_{k=1}^{K}\beta_k x_k}\beta_k}{y!}$$

$$= \frac{\beta_k \theta e^{-\theta}(y\theta^{y-1} - \theta^y)}{y!}, \qquad [A.2]$$

where $\theta = e^{\sum \beta x}$.

REFERENCES

ACHEN, C. H. (1982) *Interpreting and Using Regression.* Sage University Paper series on Quantitative Applications in the Social Sciences, 07-029. Beverly Hills, CA: Sage.

AGRESTI, A. (1984) *Analysis of Ordinal Categorical Data.* New York: John Wiley.

AGRESTI, A. (1990) *Categorical Data Analysis.* New York: John Wiley.

ALDRICH, J. H., and NELSON, F. D. (1984) *Linear Probability, Logit, and Probit Models.* Sage University Paper series on Quantitative Applications in the Social Sciences, 07-045. Beverly Hills, CA: Sage.

ALLISON, P. D. (1984) *Event History Analysis: Regression for Longitudinal Event Data.* Sage University Paper series on Quantitative Applications in the Social Sciences, 07-046. Beverly Hills, CA: Sage.

AMEMIYA, T. (1975) "Qualitative models." *Annals of Economic and Social Measurement* 4: 363-372.

AMEMIYA, T. (1981) "Qualitative response models: A survey." *Journal of Economic Literature* 19: 1483-1536.

BAUM, J. A. C., and OLIVER, C. (1992) "Institutional embeddedness and the dynamics of organizational populations." *American Sociological Review* 57: 540-559.

BEN-AKIVA, M., and LERMAN, S. R. (1985) *Discrete Choice Analysis: Theory and Application to Travel Demand.* Cambridge: MIT Press.

BERRY, W. D. (1993) *Understanding Regression Assumptions.* Sage University Paper series on Quantitative Applications in the Social Sciences, 07-092. Newbury Park, CA: Sage.

BERRY, W. D., and FELDMAN, S. (1985) *Multiple Regression in Practice.* Sage University Paper series on Quantitative Applications in the Social Sciences, 07-050. Beverly Hills, CA: Sage.

BISHOP, Y. M. M., FIENBERG, S. E., and HOLLAND, P. W. (1975) *Discrete Multivariate Analysis: Theory and Practice.* Cambridge: MIT Press.

CLOGG, C. C. (1982) "Some models for the analysis of association in multiway cross-classification having ordered categories." *Journal of the American Statistical Association* 77: 803-815.

COX, D. R., and SNELL, E. J. (1989) *The Analysis of Binary Data* (2nd ed.). London: Chapman & Hall.

CRAGG, J. G., and UHLER, R. (1975) "The demand for automobiles." *Canadian Journal of Economics* 3: 386-406.

DeMARIS, A. (1990) "Interpreting logistic regression results: A critical commentary." *Journal of Marriage and the Family* 52: 271-276.

DeMARIS, A. (1992) *Logit Modeling.* Sage University Paper series on Quantitative Applications in the Social Sciences, 07-086. Newbury Park, CA: Sage.

DeMARIS, A. (1993) "Odds versus probabilities in logit equations: A reply to Roncek." *Social Forces* 71: 1057-1065.

DOBSON, A. J. (1990) *An Introduction to Generalized Linear Models.* London: Chapman & Hall.

FIENBERG, S. E. (1980) *The Analysis of Cross-Classified Categorical Data* (2nd ed.). Cambridge: MIT Press.

FOX, J. (1987) "Effect displays for generalized linear models," in C. C. Clogg (ed.), *Sociological Methodology 1987*, pp. 347-361. San Francisco: Jossey-Bass.

FOX, J. (1991) *Regression Diagnostics: An Introduction*. Sage University Paper series on Quantitative Applications in the Social Sciences, 07-079. Newbury Park, CA: Sage.

FURSTENBERG, F., MORGAN, S. P., MOORE, K., and PETERSON, J. (1987) "Race differences in the timing of adolescent intercourse." *American Sociological Review* 52: 511-518.

GOLDBERGER, A. S. (1964) *Econometric Theory*. New York: John Wiley.

GOODMAN, L. A. (1991) "Measures, models, and graphical displays in the analysis of cross-classified data." *Journal of the American Statistical Association* 86: 1085-1111.

GREENE, W. H. (1990) *Econometric Analysis*. New York: Macmillan.

GREENE, W. H. (1993) *Econometric Analysis* (2nd ed.). New York: Macmillan.

GROGGER, J. (1990) "The deterrent effect of capital punishment: An analysis of daily homicide counts." *Journal of American Statistical Association* 85: 295-303.

HANUSHEK, E. A., and JACKSON, J. E. (1977) *Statistical Methods for Social Scientists*. New York: Academic Press.

HARDY, M. (1993) *Regression With Dummy Variables*. Sage University Paper series on Quantitative Applications in the Social Sciences, 07-093. Newbury Park, CA: Sage.

HENSHER, D. A. (1979) "Individual choice modelling with discrete commodities: Theory and application to the Tasman Bridge re-opening." *Economic Record* 50: 243-261.

HENSHER, D. A. (1981) "A practical concern about the relevance of alternative-specific constants for new alternatives in simple logit models." *Transportation Research* 15B: 407-410.

HOFFMAN, L. D., and BRADLEY, G. L. (1989) *Calculus for Business, Economics, and the Social and Life Sciences* (4th ed.). New York: McGraw-Hill.

HOFFMAN, S. D., and DUNCAN, G. J. (1988) "Multinomial and conditional logit discrete-choice models in demography." *Demography* 25: 415-427.

JACCARD, J., TURRISI, R., and WAN, C. K. (1990) *Interaction Effects in Multiple Regression*. Sage University Paper series on Quantitative Applications in the Social Sciences, 07-072. Newbury Park, CA: Sage.

KING, G. (1987) "Presidential appointments to the Supreme Court: Adding systematic explanation to probabilistic description." *American Politics Quarterly* 15: 373-386.

KING, G. (1988) "Statistical models for political science event counts: Bias in conventional procedures and evidence for the exponential Poisson regression model." *American Journal of Political Science* 32: 838-863.

KING, G. (1989a) "Event count models for international relations: Generalizations and applications." *International Studies Quarterly* 33: 123-147.

KING, G. (1989b) "A seemingly unrelated Poisson regression model." *Sociological Methods and Research* 17: 235-255.

KING, G. (1989c) "Variance specification in event count models: From restrictive assumptions to a generalized estimator." *American Journal of Political Science* 33: 762-784.

KNOKE, D., and BURKE, P. J. (1980) *Log-Linear Models*. Sage University Paper series on Quantitative Applications in the Social Sciences, 07-020. Beverly Hills, CA: Sage.

LAWLESS, J. F. (1982) *Statistical Models and Methods for Lifetime Data*. New York: John Wiley.

LEWIS-BECK, M. S. (1980) *Applied Regression: An Introduction*. Sage University Paper series on Quantitative Applications in the Social Sciences, 07-022. Beverly Hills, CA: Sage.

LIAO, T. F. (1993, August) "Confidence intervals for predicted probabilities from generalized linear models." Paper presented at the Annual Meeting of the American Sociological Association, Miami Beach, FL.

LIAO, T. F., and STEVENS, G. (forthcoming) "Spouses, homogamy, and social networks." *Social Forces*.

MADDALA, G. S. (1983) *Limited Dependent and Qualitative Variables in Econometrics*. Cambridge: Cambridge University Press.

McCULLAGH, P., and NELDER, J. A. (1989) *Generalized Linear Models* (2nd ed.). London: Chapman & Hall.

McFADDEN, D. (1974) "Conditional logit analysis of qualitative choice behavior," in P. Zarembka (ed.), *Frontiers in Econometrics*, pp. 105-142. New York: Academic Press.

McFADDEN, D. (1976) "Quantal choice analysis: A survey." *Annals of Economic and Social Measurement* 5/4: 363-390.

McFADDEN, D. (1982) "Qualitative response models," in W. Hildebrand (ed.), *Advances in Econometrics*, pp. 1-37. Cambridge: Cambridge University Press.

McKELVEY, R. D., and ZAVOINA, W. (1975) "A statistical model for the analysis of ordinal level dependent variables." *Journal of Mathematical Sociology* 4: 103-120.

MORGAN, S. P., and TEACHMAN, J. D. (1988) "Logistic regression: Description, examples, and comparisons." *Journal of Marriage and the Family* 50: 929-936.

PLOTNICK, R. D. (1992) "The effects of attitudes on teenage premarital pregnancy and its resolution." *American Sociological Review* 57: 800-811.

RINDFUSS, R. R., and LIAO, F. (1988) "Medical and contraceptive reasons for sterilization in the United States." *Studies in Family Planning* 19: 370-380.

RONCEK, D. W. (1991) "Using logit coefficients to obtain the effects of independent variables on changes in probabilities." *Social Forces* 70: 509-518.

RONCEK, D. W. (1993) "When will they ever learn that first derivatives identify the effects of continuous independent variables or 'Officer, you can't give me a ticket, I wasn't speeding for an entire hour'." *Social Forces* 71: 1067-1078.

SANTER, T. J., and DUFFY, D. E. (1989) *The Statistical Analysis of Discrete Data*. New York: Springer.

SCHROEDER, L. D., SJOQUIST, D. L., and STEPHAN, P. E. (1986) *Understanding Regression Analysis*. Sage University Paper series on Quantitative Applications in the Social Sciences, 07-057. Beverly Hills, CA: Sage.

TRAIN, K. (1986) *Qualitative Choice Analysis: Theory, Econometrics, and an Application to Automobile Demand*. Cambridge: MIT Press.

ULMER, S. S. (1982) "Supreme Court appointments as a Poisson distribution." *American Journal of Political Science* 26: 113-116.

WRIGLEY, N. (1985) *Categorical Data Analysis for Geographers and Environmental Scientists*. London: Longman.

ABOUT THE AUTHOR

TIM FUTING LIAO is Assistant Professor of Sociology at the University of Illinois–Urbana, where he teaches social statistics and demography. He holds a Ph.D. in sociology from the University of North Carolina at Chapel Hill. He publishes primarily in the areas of methodology and demography.

Quantitative Applications in the Social Sciences

A SAGE UNIVERSITY PAPERS SERIES

$9.50 each

Place
Stamp
here

SAGE PUBLICATIONS, INC.
P.O. BOX 5084
THOUSAND OAKS, CALIFORNIA 91359-9924